The Life and Death of Smallpox

This is an engaging and fascinating story of a conditional human success story. Smallpox has been one of the most devastating scourges of humanity throughout recorded history, and it is the only human illness to have been eradicated, though polio may soon follow it to official extinction through human agency. However, while smallpox is officially extinct in nature, our fears that stocks of smallpox virus may return as a weapon of bioterrorists have led to the stockpiling of vaccine, and continuing vigilance, even though the official victory over smallpox is now 15 years old. *The Life and Death of Smallpox* presents the entire engaging history of our struggle and ultimate (?) victory over one of our oldest and worst enemies. The story of the campaign to track down and eradicate the virus, throughout the world – the difficulties, setbacks, and the challenges successfully met – is a highlight of a fascinating book, but we cannot be confident of the ending. The final chapter of the book clearly and authoritatively explains the current status of the threat from the deliberate release of smallpox or other potential agents of biological terrorism.

Ian Glynn is Professor Emeritus of Physiology at Cambridge University and Fellow of Trinity College. He is the author of *An Anatomy of Thought* (1999). Jenifer Glynn is a Cambridge historian and author of *Tidings From Zion* (2000).

The Life and Death of Smallpox

IAN AND JENIFER GLYNN

CAMBRIDGE
UNIVERSITY PRESS

PUBLISHED BY CAMBRIDGE UNIVERSITY PRESS
40 West 20th Street, New York, NY 10011-4211, USA

http://www.cambridge.org

© Ian and Jenifer Glynn 2004

First published in the United States of America and Mexico by
Cambridge University Press 2004, under license from Profile Books

Printed in the United States of America

Typeset in Poliphilus

A catalogue record for this book is available from the British Library

ISBN 0 521 84542 4 hardback

Contents

List of illustrations

Acknowledgements

Writing the biography of a subject whose known life extended through three millenia and embraced five continents is a wide-ranging task, and we want to thank the many people whose help has made that task a pleasure. Graeme Mitchison, Douglas and Clare Fearon, Judith Glynn, Alan Glynn, Colin Franklin and Eckard Wimmer have read and criticised the entire text. Geoffrey Smith and Ian Ramshaw have given valuable advice on virology, Peter Lachmann on immunology, John Twigg on history. Seemingly endless questions have been answered by colleagues at Cambridge – Arnold Browne, Roger Dawe, John Easterling, Eric Handley, Simon Keynes, Sachiko Kusakawa, John Lonsdale, William St Clair, Peter Sarris. For errors that have survived this flood of advice, we accept responsibility.

The World Health Organization's massive 1988 publication *Smallpox and its Eradication* (by Frank Fenner, Donald A. Henderson, Isao Arita, Zdenek Jezek, and Ivan D. Ladnyi) has been a mine of information, and we are very grateful to the Organization for generous permission to reproduce many illustrations. Laura Cordy has been a wonderful help in preparing these and other illustrations for publication.

We would particularly like to thank the director of the Jenner Museum in Berkeley, Gloucestershire, and the librarians of the University Library, the Geography Library, the Medical Library, the Needham Institute, and Trinity College Library, in Cambridge, and of the Royal Society of Medicine in London.

Finally we want to thank our agent Felicity Bryan, our editor Penny Daniel, and our publisher Andrew Franklin, for encouragement, advice and understanding during the successive stages of the book's production.

Ian and Jenifer Glynn
Cambridge

1

'The most terrible of all the ministers of death'

Te Deum Laudamus!
Little Wolfgang has got over the smallpox safely!
Letter from Leopold Mozart, in Vienna, 10 November 1767

Mozart was eleven when he became delirious with smallpox, and he was lucky to survive with only a few pockmarks; in England in 1700 the disease had killed Queen Anne's eleven-year-old son, ending the Stuart line. It was a terrible century for smallpox all over the world, from the royal families of Austria, Spain and France to the isolated communities of Iceland and Greenland, from the long established centres of the disease in China and India to the newly colonised lands of the New World. The growing towns were particularly vulnerable – most children born in eighteenth-century London had smallpox before they were seven, and more than 6,000 people died in the epidemics in Rome in 1746 and 1754.

There was no effective treatment. It flourished for many centuries, bringing horrific death, disfigurement and blindness on a huge scale; armies were checked, populations decimated, economies ruined. It is only in the last few years that drugs able to attack the smallpox virus have begun to emerge. They should not be needed, for a generation ago smallpox was defeated by a vigorous world-wide campaign of vaccination — the first (and so far the only) serious infectious disease to be totally eradicated. This book tells the story of smallpox and of the long battle to control and ultimately eliminate it. The new horror is that there has to be a postscript — the threat that bio-terrorists may try to exploit our unvaccinated state and use smallpox as a weapon.

Like most people over thirty-five we, the authors, both have pock-marks on our upper arms resulting from vaccination against small-pox when we were babies. But routine vaccination of civilians in the United Kingdom and North America stopped in the early 1970s, and world-wide by 1980; and so effective has the eradication of smallpox been, that most people alive now have never seen a case.

So what was it like? As children, we were told it was like chickenpox but worse. In fact it is not related to chickenpox, and it was unimag-inably worse. In an unvaccinated population, something like 10–30 per cent of all patients with smallpox would be expected to die. And dying was not easy; smallpox was, as Macaulay wrote, 'the most ter-rible of all the ministers of death'.

Smallpox is virtually restricted to humans; patients are highly infectious for only about two weeks, and those who recover from it are immune for the rest of their lives. In small isolated communities the disease was therefore likely to die out for lack of sufficient potential victims. In larger communities, even if they were isolated, there would always be sufficient susceptible children to keep the infection going, though the number infected might fluctuate wildly, with peaks (epi-demics) only every few years, each followed by a trough until the birth of more children restored the supply. In such a community, smallpox was said to be endemic, and the majority of its victims were children. In communities never previously exposed, or where the disease had been absent for many decades, when it arrived the whole population was at risk.

Whether child or adult, about twelve days after infection the patient was struck with a sudden fever, splitting headache and often backache, and sometimes vomiting. Two to three days later, the rash appeared, the temperature fell, and the patient, who had been feeling extremely

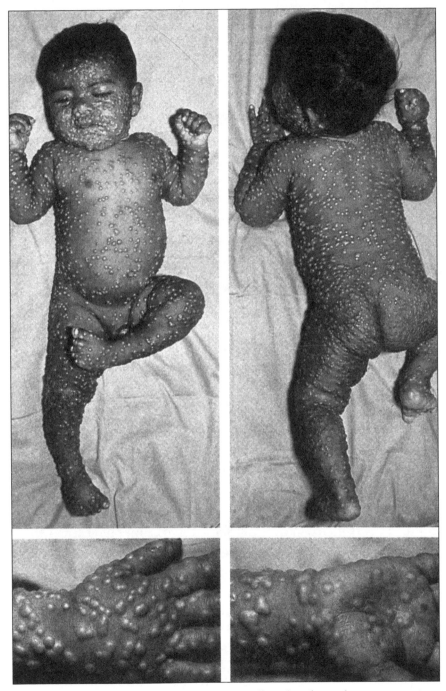

1. A child with smallpox; the 'recognition card' used in the eradication campaign

ill, felt a little better. Over the next two weeks the fever returned, and the rash went through a characteristic series of changes. Starting flat, the spots gradually became raised above the skin surface and felt hard, like embedded lead shot. They then softened, becoming filled first with clear fluid and then with pus. In most cases the 'pustules' gradually flattened, dried up and scabbed, the fever went, and within three weeks of the appearance of the rash the scabs fell off, often leaving permanent, pitted scars – the pockmarks – particularly on the face.

Usually, the rash began in the lining of the mouth and throat and on the face, then spread to the upper parts of the limbs and to the trunk, and finally to the hands and feet. In mild cases the rash was sparse and the scarring very limited. Lady Mary Sidney, who had nursed Elizabeth I of England through a frightening but not too disfiguring attack, was less lucky herself. 'I left her a full fair lady, in mine eyes at least the fairest,' wrote her husband sadly, 'and when I returned I found her as foul a lady as the smallpox could make her.'[1] In severe cases, the pustules could be so crowded that they fused together ('confluent smallpox'), and the number dying rose to about 60 per cent; in the rare and most horrific form there was severe bleeding ('haemorrhagic smallpox') and over 90 per cent died. If the rash was extensive in the mouth and throat, eating and drinking could become intolerably painful, and if the eyes were affected the patient could be permanently blinded. At the end of the eighteenth century about a third of all cases of blindness in Europe are thought to have been the result of smallpox; and the death toll was terrible – about 400,000 a year in Europe, excluding Russia. It was children who suffered most – in English towns, nine out of ten of those dying from smallpox were children under five years old.[2]

Since smallpox cannot be treated effectively, prevention is crucial. Inoculation, which was introduced to Europe at the beginning of the eighteenth century from far earlier origins in the East, first gave the idea that prevention might be possible. It involved the insertion of a small amount of matter from the pustule of a smallpox patient into the skin of a healthy subject. If it worked, as it generally did, it produced a very mild attack of the disease, and gave lasting immunity. But during this mild attack the inoculated subjects were themselves infectious and so the disease could spread further; and occasionally the disease produced by inoculation was not mild, and might even be fatal.

Edward Jenner, at the end of the eighteenth century, took the vital next step. By introducing 'vaccination' (that is, inoculation with material not from smallpox pustules but from the pustules of cows suffering from the related disease cowpox), he usually produced a milder and uninfectious reaction which gave immunity though, as it turned out, not quite permanent immunity. This drawback was easily tackled by later revaccination; and the idea of eradication changed from a utopian dream to a practical possibility – though it took nearly 200 years to achieve.

By 1967, when the World Health Organisation (WHO) started its final eradication campaign, smallpox had virtually disappeared from Europe and the Americas, but from figures reported to the WHO's Smallpox Eradication Unit it is now reckoned that the disease was still endemic in thirty-three countries, including the whole of the Indian subcontinent, and that there were 10 to 15 million cases and about 2 million deaths per year world-wide.[3] Twelve years later, after an amazing effort of organisation and co-operation the world was officially declared free of smallpox.

2

From myths to mummies

The scorching sands of Afric gave him birth
Thence sprang the Fiend and scourged the afflicted earth
William Lipscomb, 'The Beneficial Effects of Inoculation', the Chancellor's prize poem, Oxford, 1772

William Lipscomb, Oxford undergraduate and son of a Winchester surgeon, may well have been right when he claimed that smallpox originated in Africa – though it is more likely that it emerged in a populous river valley than in the scorching sands. Myths and theories abound. Some tried to link smallpox with biblical diseases; Philo of Alexandria, the Jewish emissary to Caligula's Rome in the first century AD, confidently described the sixth Egyptian plague ('boils') as a red eruption in which 'the pustules, confluent into a mass, were spread over the body and limbs' – a good description of confluent smallpox, but a creative interpretation of the biblical verses, and one which ignores the inconvenient point that the biblical plague affected beasts as well as people.[1] The Revd Edmund Massey, preaching against inoculation in eighteenth-century London, saw smallpox as a trial or punishment sent by God, as in Job's boils.[2] In the eighteenth century too, Father Pierre Cibot, a Jesuit missionary studying ancient medical texts in Peking, claimed that a disease resembling smallpox had existed in China for 3,000 years – a claim often repeated but not now widely accepted.[3] Another suggestion was that smallpox entered China with the Huns, arriving from the north about 250 BC, before the Great Wall was built.[4]

A disease referred to as *masurika* is mentioned in two Indian med‚ ical texts thought to be about 2,000 years old. The word is derived from the name of an orange lentil whose shape and colour are sup‚ posed to resemble the pustules of smallpox, and the same word is used much later to refer to what was almost certainly smallpox.[5] So it might seem that smallpox existed in India 2,000 years ago. But in 1981 Ralph Nicholas, an anthropologist from Chicago, pointed out that neither of the two early texts appears to take masurika very seriously.[6] One does not include it in an enumeration of diseases, and mentions it only briefly in a section on treatment; the other gives a short description that fits smallpox well enough except that there is no mention of its life‚threatening quality or epidemic tendency, and the description is included in a section on minor ailments, sand‚ wiched between a discussion of premature greying of the hair and the treatment of congenital moles and freckles.[7] It is not until the seventh century AD that Indian medical texts describe a fatal disease with the features of smallpox. So either the disease changed, or, more probably, the early disease was not smallpox.[8]

There is, though, a long‚held assumption that Indians have wor‚ shipped a goddess of smallpox for two or three thousand years. In 1767 John Zephaniah Holwell[9] gave a talk to the College of Physicians in London. He had spent nearly thirty years of his life in India and he had been a surgeon in the East India Company, a member of the coun‚ cil running the company's affairs, the man called on to take charge when the Nawab of Bengal attacked the British settlement in Calcutta (and the governor and many of the senior officers skedaddled down the Hooghly), and one of the few survivors of the subsequent incarcera‚ tion in the Black Hole – the infamous jail of Calcutta's Fort St Wil‚ liam. For a short time, too, he had been deputy governor of Bengal. But Holwell did not talk to the physicians about his life story. In an address mainly concerned with the manner of inoculating for the smallpox in the East Indies, he told them that the *Atharva Veda* – the fourth book

2. *Sitala, Hindu goddess of smallpox*

of the ancient Hindu scriptures, written according to the Brahmins more than 3,000 years ago – 'instituted a form of divine worship, with Poojahs or offerings, to a female divinity, stiled by the common people … the goddess of spots …' This story was often repeated and widely believed, until it lost credibility at the beginning of the twentieth century when Sanskrit scholars said that they could not find any such reference in the ancient scriptures.[10]

It is true, though, that for a long time a Hindu goddess of small-pox has been worshipped with much enthusiasm throughout India, and that over the whole of northern and central India and Nepal she is known by a single name, Sitala – literally 'the cool one'.[11] She is believed to be able both to bring smallpox and to help those suffering from it, and she is represented in different forms. She may appear deceptively charming, sitting decorously side-saddle on a donkey, carrying a water pot and a broom, with a winnowing fan on her head, and dressed in red with polka dots. Or she may stand threat-eningly with crooked daggers in both hands raised to strike. But does she provide evidence for the existence of smallpox in ancient India? References to her are not found earlier than the beginning of the sixteenth century and, though sculptures with variations on the same theme have been found in the thirteenth, twelfth and ninth centuries, there is no solid evidence that she existed at still earlier times; nor is it certain that she has always been connected exclusively with small-pox – measles and chickenpox probably came into the picture too.[12] Smallpox gods or goddesses have also been found in China, Japan, West Africa and Brazil, but though they too are often assumed to go back to ancient times it is not clear that they do.

More convincing, though still uncertain, evidence that smallpox flourished in the first millennium BC is the plague of Athens described by Thucydides in lurid detail in his *History of the Peloponnesian War*.[13] The epidemic started in 430 BC and by its end had killed a quarter of the Athenian army, as well as Pericles and all his legitimate children. Whatever it was, it began in Ethiopia

> and then descended into Egypt and Libya and spread over the greater part of the King's territory. Then it suddenly fell upon the city of Athens, and attacked first the inhabitants of the Peiraeus, so that the people there even said that the Peloponnesians had put poison into their cisterns; for there were as yet no fountains there. But afterwards it

reached the upper city also, and from that time the mortality became much greater ...[14]

And Thucydides continues:

> *... I shall describe its actual course, explaining the symptoms, from the study of which a person should be best able ... to recognize it if it should ever break out again. For I had the disease myself and saw others sick of it.*

What is striking about Thucydides' account is that it includes most of the features characteristic of smallpox – the infectiousness, the immunity conferred by an attack, the sudden onset, headache, 'inflammation of the eyes and the parts inside the mouth', the skin 'livid and breaking out in small blisters and ulcers', the eruption starting with the head and moving to the extremities where it 'attacked the privates and fingers and toes', restlessness, vomiting, and in some patients diarrhoea, convulsions or blindness. But it fails to mention residual pockmarks, and it also includes features not usually associated with smallpox, such as the loss of fingers or toes, and the loss of memory or of the ability to recognise people and objects. The loss of fingers and toes could be the result of gangrene, which though not common in smallpox is not unknown. Alternatively, the Greek words can refer not only to 'loss' but also to 'loss of the use of', and such loss might be caused by pustules crowded on the digits, with or without secondary infection. Loss of memory and of the powers of recognition could be signs of the encephalitis that is occasionally a complication of smallpox. And there is, of course, no reason to assume that every feature described by Thucydides was the result of whatever it was that caused the epidemic. Epidemics have no monopoly, and in any epidemic there will be many patients suffering from the epidemic disease who also have other diseases with their associated symptoms. Physicians who have wondered about the cause of the plague of Athens have generally regarded measles, plague, typhus and smallpox as the candidates, with smallpox the favourite and typhus the runner-up. Hans Zinsser, who in his classic *Rats, Lice and History*[15] argued strongly for smallpox, pointed out that less than forty years after the plague in Athens a similarly smallpox-like epidemic attacked the Carthaginian army besieging Syracuse – as described by Diodorus Siculus, though he was writing more than 300 years after the event. (A result of this epidemic was that the Carthaginians failed to gain complete control of Sicily – a failure which, Zinsser likes to think, tipped the balance against Carthage in the First Punic War that was to follow more than a century later.)

The belief that smallpox existed more than 2,000 years ago does not, though, depend solely on historical accounts such as those of Thucydides or Diodorus, or on doubtful interpretations of ancient Chinese or Indian medical texts; it is also supported by the study of Egyptian mummies.

Marc Armand Ruffer, son of Baron Alphonse Jacques de Ruffer, a banker of Lyons, was educated at Oxford, and then took a medical degree in London.[16] Returning to France, he became a pupil of Pasteur and Metchnikoff at the Institut Pasteur, and then, in 1891, came back to London as the first director of the British Institute of Preventive Medicine – the body that would eventually become the Lister Institute. While testing new antisera to diphtheria, he was so severely paralysed by the diphtheria toxin that he felt he must resign the directorship. He went to Egypt to recuperate, settled in Ramleh, became Professor of Bacteriology at the Cairo School of Medicine, and played a crucial part in ridding Egypt of cholera by rigorous enforcement of quarantine stations along the routes of pilgrimage. But his interests were not limited to current problems; he was fascinated by ancient Egypt, writing on topics ranging from the way shepherds made bread from millet seed to the way the Ptolemies monopolised the (vegetable) oil trade – forbidding imports, registering oil presses, searching for concealed presses and imposing heavy fines when they were found.[17] Linking his medical and antiquarian interests, he was among the first to look at mummies for signs of disease.

In 1911, in collaboration with the Professor of Pathology at Cairo,

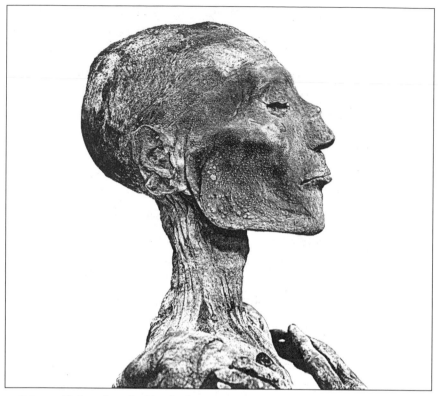

3. Mummified pockmarked head of Ramses V

A. R. Ferguson, he published a short account of a mummy of the twentieth dynasty (1200–1100 BC) whose 'body was the seat of a peculiar vesicular ... eruption which in form and general distribution bore a striking resemblance to that of small-pox'.[18] They had been allowed to remove a bit of skin, which when examined under the microscope showed that the vesicles 'must have originated and developed in the prickle layer, i.e. the situation in which the small-pox eruption is first seen'. Ruffer mentions finding a very similiar skin eruption in a mummy of the eighteenth dynasty (1580–1350 BC), but the most convincing example is that of Ramses V, who died in 1157 BC.[19] His mummy, which was photographed in 1912 by the anthropologist Grafton Elliott Smith, has what looks like a pustular eruption on the lower neck, face and shoulders, and also on the lower abdomen and scrotum.[20] Elliott

Smith, Ruffer and Ferguson all thought this eruption looked like smallpox. In 1979, the physician–epidemiologist Donald Hopkins got permission from the then President of Egypt, Anwar Sadat, to examine Ramses V's partially unwrapped mummy,[21] but he found that it 'was not possible to examine the palms or soles where the presence of pustules would be highly characteristic of smallpox, because [the mummy's] arms were folded across his chest with the palms down, and the shroud was stuck to his soles'.

Hopkins wanted to excise one of the pustules so that it could be looked at with an electron-microscope to see whether it contained the now well-known, characteristic, brick-shaped particles of smallpox virus, but that was not allowed; and examination of tiny pieces of tissue that had fallen on the shroud was unsuccessful. He does though provide interesting indirect support for the smallpox theory. Pharaohs were usually buried after sixty days of mummification; Ramses V was not buried until the second year of his successor's reign. Why the delay? Hopkins has a plausible answer. If Ramses V really did die of smallpox, his embalmers 'would likely have suffered a fearsome epidemic about two weeks after starting to prepare his body, and the source of such a local outbreak would surely have been suspected ... Fear of the infection (if not an acute shortage of embalmers) could then have postponed the remaining preparation and burial'.

3

Coming into focus: AD 0 to 1500

Nearly 2,000 years ago, more stories of smallpox emerged from the Middle East and the Far East.

Indirect evidence that there had been an epidemic in China in about 45 AD comes from the writings of the great 'physician and alchemist' Ko Hung.[1] His book, *Handy Therapies for Emergencies*, was finished in about 340 AD, though it now exists only in a version revised and expanded by another physician and alchemist in about 500 AD. It says:

> *Recently some people have suffered from seasonal epidemic eruptions which attack the head face and trunk. In a short time they spread all over the body. They look like fiery boils, all containing a white fluid. The pustules arise all together, and later dry up about the same time. If they are not treated immediately, many of the more severely afflicted patients will die in a few days. Those who recover are left with purplish or blackish scars, the colour of which takes years to fade ... People say it first appeared from the West in the fourth year of the Yung-hui reign-period and passed eastwards, spreading all over the country. In the mid-Chien-wu era [our soldiers] caught it when attacking marauders at Nan-yang; for this reason one of its names is still 'marauders' pox.*

The general view is that the reference to the Yung-hui period (which is seventh century) must be the result of a copyist's error, and that the relevant Chien-wu period is just before the middle of the first century AD.

In 166 AD, Rome under the Stoic emperor Marcus Aurelius Antoninus was struck by a severe epidemic that had started in Mesopotamia. Galen, the Greek physician whose views were to dominate medical thought until the Renaissance, was in Rome at the time; and though he wisely soon left, he was there long enough to give a description of the disease that points rather strongly to smallpox, and particularly to haemorrhagic smallpox.[2] Lasting about fifteen years and extending over a wide area, this 'Antonine plague' led to the abandonment of cities and villages, the postponement of military campaigns, and the deaths of a very large number of people including, at the tail end of the epidemic, Marcus Aurelius himself. At that time he and his son and heir, Commodus, were campaigning against Germanic tribes east of the Adriatic, and there is a story that the dying Marcus Aurelius, afraid that his son would catch the disease, sent him away. How much this dismissal might have helped Commodus to survive there is no way of knowing, but if it did it also ensured that for the next twelve years Rome was ruled by a debauched and tyrannical emperor. (At the age of thirty-one, Commodus was poisoned by one of his 300 concubines, and then strangled in his bath by a wrestler called Narcissus).[3]

During the next five centuries outbreaks of what was probably smallpox were reported from a variety of places in Asia, Europe and North Africa. In 302 AD, Eusebius – later Bishop of Caesarea, and the author of the first draft of the Nicene creed – described an epidemic in Syria in which a skin eruption spread over the whole body and often led to blindness or death.[4] In the middle of the fifth century the Huns, whose invasion of eastern France was being hindered by some kind of pestilence, confronted Nicaise, Bishop of Rheims, and beheaded him, along with his virgin sister, on the steps of his own cathedral. The previous year Nicaise, with the help of holy oil, had recovered from what was supposed to be smallpox – possibly the same pestilence that was hitting the invaders – and when he was canonised he became the patron saint of smallpox victims.[5] He is portrayed carrying his head because he is traditionally believed to have walked back into the cathe-dral, not falling dead until he reached the altar.

4. St Nicaise, the patron saint of smallpox, and his sister, on the north portico of Rheims cathedral

The sixth and seventh centuries provide dramatic examples of interactions between outbreaks of smallpox and the spread of religious beliefs.

There is a story that in 569 AD, or thereabouts, an outbreak of a vicious disease involving pustules on the skin was critical in defeating a siege of Mecca by an army of Christian Abyssinians coming from the Yemen – a siege often referred to as 'the elephant war' though only one elephant was involved. At that time – about the time of Muhammad's birth – Mecca was still largely pagan, worshipping a black stone (probably a meteorite), along with images of a host of tribal gods, in a sanctuary called the 'cube' (Kaaba) that was already a centre for pilgrimage (and which would later become the holiest site in Islam). A large force of Abyssinians led by Prince Abraha on a white elephant is said to have besieged the city, aiming to destroy the pagan sanctuary, but the besieging army melted away as it succumbed to the disease.[6] The sanctuary was preserved, with a greatly enhanced reputation, and its idols remained until Mecca surrendered to Muhammad and his army marching from Medina in 630. If the story is true, and if it really was smallpox that decided the outcome of the elephant war, we can say that without the influence of smallpox Muhammad would probably have grown up in a Christian rather than a mainly pagan town, and history might have been very different.

The evidence that it was smallpox is flimsy, but the diagnosis is plausible since smallpox was about both in the Middle East and in Europe. Marius, a skilled goldsmith who became Bishop of Avenches (near Lausanne) in 573, described an epidemic in southern Europe using the Latin word *variola*, from *varius* meaning spotted; this is the word that for the whole of the last millennium has been used specifically for smallpox, but it is not certain that it had the same meaning in the sixth century. There can be little doubt, though, that it was smallpox that swept through parts of France and northern Italy in about 580, and was described by Gregory, Bishop of Tours:

> ... – *such was the nature of the infirmity that a person, after being seized with a violent fever, was covered all over with vesicles and small pustules ... The vesicles were white, hard, unyielding and very painful. If the patient survived to their maturation, they broke and began to discharge, when the pain was greatly increased by the adhesion of the clothes to the body ... Among others, the Lady of Count Eborin, while labouring under this pest, was so covered with the vesicles, that neither her hands, nor feet, nor any part of the body, remained exempt, for even her eyes were wholly closed up by them.*[7]

The countess did recover though:

> *When nearly at the point of death, she received some of the water in which the tomb of the blessed saint [St Martin] had been washed at*

the Lord's Passover.* — This having been taken as a drink, and applied to her sores, the fever abated, the discharges from the vesicles were made without pain, and she was soon after healed.

Nearer to Mecca, an outbreak of pestilence in Alexandria in 622 AD was described by a Christian priest, 'Aaron the Physician', and his account (though now known only in quotations by others) sounds convincingly like smallpox.[8]

Whether or not smallpox was partly responsible for the situation that led to the creation and explosion of Islam, there is no doubt that that explosion helped to spread the disease. In communities in which smallpox is endemic it is mainly young children who are at risk, but where there is interchange between communities with different histories of exposure to smallpox, there is a greater opportunity for the virus to spread and a greater likelihood of severe epidemics. The huge movements of people that helped to disseminate Islam must have provided many such opportunities. As the eighteenth-century English physician John Haygarth put it (discussing the conquest of Alexandria by Caliph Omar in 640 AD), 'By an immense concourse of people, comprehending the besiegers and the besieged, the small-pox was widely dispersed among mankind.'[9] And, of course, the besiegers included not just soldiers but also those that Jenner's friend James Moore[†] called 'their plural wives, children and slaves'.[10]

Muhammad's revelations did not begin until he was in his forties, and the *Hegira* – the flight from Mecca to Medina that is the starting point of the Muslim calendar – was in 622, when he was about fifty-one. By the time of his death, his supporters were in control of most of the Arabian peninsula; and the tide of Muslim expansion continued to flow for the rest of the seventh century and well into the eighth. It covered Persia and parts of Turkestan to the east, Syria, Palestine, Mesopotamia and Armenia to the north, and Egypt, the rest of the north

---

* That is, the Thursday before Easter, when the events of the last evening of Christ's life, including the Last Supper, were celebrated.

† Surgeon, and brother of Sir John Moore of Corunna fame.

African coast, Spain and the southern half of France to the west. Only in 732, just a hundred years after Muhammad's death, was the tide halted, by Charles Martel at the battle of Poitiers in central France.

It would be wrong, though, to follow Haygarth in thinking of the Muslim expansion as bringing smallpox into a Europe that had until then been totally free of it. By the time the Arab armies were turned back at Poitiers they had very likely brought epidemics to the Mediterranean countries but, as we have seen, smallpox had probably already been there several times in earlier centuries.

———

Around the middle of the sixth century, the Japanese were introduced to both Buddhism and smallpox; and the acceptance of the religion was much influenced by outbreaks of the disease.[11] The introduction to Buddhism came in the form of a gift — to Kimmei, the emperor of the Japanese kingdom of Yamato — from the ruler of a kingdom in southwest Korea, who hoped to secure Japanese support in dealing with the military attacks and political machinations of his troublesome neighbours. The gift consisted of a copper and gold image of Buddha, flags, umbrellas and several volumes of the Sutras, together with a memorial from the ruler extolling the virtues and efficacy of the new religion. And it presented the emperor with a difficulty: ought the new image to be worshipped or not?

Kimmei's 'great-minister', who was, incidentally, also the father of two of Kimmei's secondary wives,* argued that since 'all the western frontier lands without exception' worship Buddha it would be absurd for Yamato to refuse to do so. Another senior minister (the 'head of the clans'), together with the chief priest, pointed out that the worship of foreign gods instead of the traditional national ones was bound to incur the wrath of those that were neglected. Diplomatically, Kimmei gave the image to his great-minister, with instructions to worship it, but otherwise allowed the worship of the traditional gods to continue.

---

*Or possibly father of one and brother of the other.

This solved Kimmei's immediate problem. But when an epidemic of smallpox broke out and persisted, the head of the clans and the chief priest attributed this to the anger of the traditional gods; they persuaded Kimmei to order the image to be thrown into the Naniha canal and the temple which the great-minister had built for it to be burnt. The origin of the smallpox epidemic is uncertain, but it is likely to have come from either Korea or China.

That might have been the end of the story, but twenty-five years after the gift of the ill-fated image, the ruler of the south Korean kingdom, still seeking Japanese help, sent a new image of Buddha, together with monks, nuns, a reciter of mantras, a temple architect and 200 volumes of Sutras. And seven years later he sent two further images. By this time, Kimmei had died and been succeeded by his second son; and the great-minister had also died and been succeeded by one of his sons. With this change in cast, there was a re-run of the old story.

The images, the monks and the nuns had been put under the care of the new great-minister, who had erected a temple for them. A new outbreak of smallpox was attributed by the traditionalists to the neglect of the traditional gods; Buddhism was proscribed, the temple was burnt, the remains of the images were flung into the same canal, and the nuns were stripped and flogged near the market place. But this time there was a sequel.

When, in spite of all these efforts the epidemic persisted, there was a new doubt: might it be a punishment for the destruction of the Buddhist images, rather than a result of the anger of the traditional gods? The emperor decided to allow a limited worship of Buddha, while still forbidding attempts at conversion. The following year (586), the emperor himself died of smallpox and the new ruler was his half-brother, an older son of Kimmei. Two years later, this emperor too was dead, but before dying he had become the first emperor to embrace Buddhism. In spreading Buddhism, the new great-minister had succeeded where his father had failed, and by the time he himself died, nearly forty years later, there were in Yamato nearly fifty Buddhist temples, with over 800 priests and over 500 nuns.[12]

Even this is not the end of the connection between Buddhism and smallpox in Japan. The scattered distribution of the population at that time prevented smallpox from becoming endemic, but twenty-five years after the founding of Nara (early in the eighth century) there was a new epidemic, supposedly caused by a fisherman who had returned to Kyushu after becoming infected while shipwrecked in Korea.[13] Again arguments arose between the Buddhists and the supporters of the Shinto gods, and it is smallpox we must thank for the erection, later in the eighth century, of the huge bronze Buddha in Nara, whose out-stretched hand is said to be able to support the weight of three men. At the other end of the scale, the Empress Koken ordered a million tiny pagodas, each a few inches high, containing charms to drive out disease – the earliest block printing on paper in Japan; she died, probably of smallpox, just as they were completed.[14]

One of the consequences of the Arab Islamic explosion in the Middle East was the flourishing of what is generally known as Arabian medi-cine, though its practitioners were not all Arabs (many of them were Persian) or all Muslim (some of them were Jews or Christians). In the eighth and ninth centuries, many of the books associated with the school of Hippocrates or with Galen were translated into Arabic but, by the start of the tenth century, medical scholars writing in Arabic were beginning to put less effort into translating classical Greek texts and more into original scholarship. One of the most interesting of these scholars was Abu Bakr Muhammad ibn Zakariyya al-Razi – gener-ally known in the West as Rhazes.[15] Born in Rayy, Persia, in 854 AD, he spent most of his life in Baghdad, where he directed a hospital. He was very critical of religion and of religious prophets; men such as Euclid and Hippocrates were, he felt, more useful. He said that a man of sci-ence who knew the work of his predecessors had, because of this, an advantage over them (however eminent they were) and could proceed with new discoveries – an early version of the point made by Newton about standing on the shoulders of giants. Of his voluminous medical works, the most relevant here is his *Treatise on the Smallpox and Measles*, which provides much the most detailed description of smallpox written before modern times, and was the first to distinguish clearly between smallpox and measles. He noted that the disease was seasonal, breaking out particularly in the spring, and that it was predominantly a disease of children. We can safely assume that the reason it was a disease of children in Baghdad is that it was endemic there, so that most adults had had the disease and were now immune. But notions of infection and immunity were not available to Rhazes, and he was not one to explain things in terms of punishment by angry gods; so how could the prevalence among children be explained?

Since the disease affected nearly all children, it could be thought of as the result of something innate in each child. And the nature of the disease, with matter being discharged from pustules, suggested that something undesirable was being got rid of through the skin. Rhazes had an ingenious theory, based on notions dating back to the Greeks, that 'everyone from the time of his birth till he arrives at old age is continually tending to dryness', and that the blood of infants and young children is both moister and warmer than that of adults, and still more than that of old men. 'For this reason', he argued

> ... the Small-Pox arises when the blood putrefies and ferments, so that the superflous vapours are thrown out of it, and it is changed from the blood of infants, which is like must [i.e. unfermented or incompletely fermented grape juice], into the blood of young men, which is like wine perfectly ripened: and the Small-Pox itself may be compared to the fermentation and hissing noise which takes place in must at that time. And this is the reason why children ... rarely escape being seized with this disease, because it is impossible to prevent the blood's changing from this state into its second state, just as it is impossible to prevent must (whose nature is to make a hissing noise and to ferment) from changing into the state which happens to it after its making a hissing noise and its fermentation.[16]

A rather different version of the theory that the cause of smallpox is innate within the child is associated with a contemporary of Rhazes called Isaac ben Solomon Israeli, more often referred to as Isaac Judae-us. Born in Egypt, in middle life he emigrated to Tunisia where he became physician to two successive rulers, wrote books on philosophy and on medicine, and had a reputation for wit. James Moore, writing in 1815, explains Isaac's theory:

> *He supposed that the foetus in the womb was tainted with some portion of this noxious female fluid [the mother's menstrual blood, regarded as unclean in the Old Testament]; which being unfit for nutrition, was thrown by nature into certain places near the skin, lest it should injure the principal organs. After the birth of the child, the morbid humour remained quiet, until it was set in commotion by some external cause, such as bad food, or corrupt air, when it was expelled from the surface of the body, in the form of Small Pox, which he considered to be a fortunate ejectment.*
>
> *Thus did this uninspired Jew cast the reproach of the Small Pox, like another original sin, upon women. And perhaps it was owing to the Mahometans not entertaining due respect towards the sex that these indelicate hypotheses were admitted by them. But it is astonishing, that they were also credited under various modifications, by many of the most celebrated Christian physicians, down to the eighteenth century.[17]*

And having had his fun at the expense of physicians of the three mono-theistic faiths, Moore passes on to other topics. But the notion that the expulsion of a 'womb poison' accounted for eruptive diseases in chil-dren is also found in twelfth-century Chinese writings on children's diseases,[18] where it is unlikely to have been derived from Old Testament prejudices.

For medieval patients as for modern, a physician's notions about pathology were of less immediate concern than his notions about treat-ment. Again, Rhazes was the first to give a systematic and detailed description of the way he would treat smallpox at each stage of the disease. Where there is no known effective treatment, deciding between alternatives must depend partly on local tradition and partly on the theories held by the physician. The Arabian physicians inherited from the Greeks a firm (though totally unjustified) belief that many diseases were the result of imbalance between the four 'humours'— blood, phlegm, yellow bile and black bile — so an important aim of treatment was to redress the balance. It is this belief — or elaborations on it — that accounts for the enthusiasm for blood-letting and purging that was so marked a feature of medicine from classical times to the nineteenth century. Rhazes recommended bleeding those at risk of smallpox if the weather was unpropitious and the patient was 'moist, pale and fleshy', or 'well-coloured and ruddy', or 'swarthy and loaded with flesh'. 'A vein may be opened in those that have reached the age of fourteen years; and cupping glasses must be applied to those that are younger ...' When the early symptoms of smallpox appeared – fever, violent headache, backache, redness of the eyes – Rhazes advocated bleeding 'even to the point of fainting'; at this early stage the patient should drink 'water cooled with snow', and even 'wash himself in cold water, and go into it, and swim about in it'. But once the rash began to appear Rhazes' aim changed to accelerating the eruption, and for this he recommended heat treatment:

> Let the patient put on a double shirt, with the upper border closely buttoned; and underneath let there be placed two small basins of boiling water, one before and the other behind him, so the vapour may come to the whole body except the face; and the skin may be rarefied, and disposed to receive and evaporate the superfluous humours.[19]

Sweating was supposed to be beneficial, provided that it did not lead to fainting. If it did, it was to be stopped immediately. If the rash began to affect the eyes, he had a battery of remedies ranging from rose water in which sumach had been macerated to 'Nabathaean caviare in which there is no vinegar, nor any other acid.' (Just what, if anything, the effective constituent of Nabathean caviare was, is obscure, but At-Tabari, one of Rhazes' teachers, says that an eye-lotion made with caviare 'strengthens and preserves the pupil and does away with any opacity'.[20])

Rhazes also provided immensely detailed dietary instructions, some of which are surprising:

> *As to melons, especially sweet ones, they are entirely forbidden; and if the patient happens to take any he should drink immediately after it the inspissated juices of some of the acid fruits.*[21]

Hali Abbas, another Persian physician, who was born a few years after Rhazes' death, recommended puncturing the smallpox pustules and rubbing them with an ointment containing salt – a procedure that must have been extremely painful and that led Moore to remark, 'How singular! that almost every attempt made by these learned men to do good, must have done mischief'.[22] (Abbas did, though, make a significant contribution to our understanding of smallpox when he pointed out that the disease tended to occur if the patient had been in the vicinity of persons affected with smallpox, or had been breathing air contaminated with the vapour from smallpox pustules.[23])

Bleeding, purging, sweating, puncturing pustules, the use of astringent eye-drops, and modifying the patient's diet, though generally ineffective and sometimes harmful, were all treatments with some sort of rational basis, even if the premises on which they were based were unsound. But there were some treatments that lacked even that fig leaf. Avicenna, the most famous Persian physician of the eleventh century, prescribed the usual bleeding, sweating and decoctions to expel the 'morbid humours', but he also recommended that pustules be opened on the seventh day with golden needles.[24] This custom survived in fourteenth-century England, where it was used by John of Gaddesden, prebendary of St Paul's Cathedral, Fellow of Merton College, Oxford, and possibly the original model for Chaucer's 'Doctor of Physic'[25], who

... knew the cause of every malady,
Were it of cold, or hot, or moist, or dry,
And where engender'd, and of what humoúr

Certainly John didn't lack confidence. He was the author of a treatise on medicine, which he chose to call *Rosa Medicinae* – because, he
explained, as the rose has five sepals, so the book has five parts; and as
the rose excels all flowers, so the book excels all treatises on the practice
of medicine.[26] In it he mentions using another improbable treatment
for smallpox – the red treatment[27] – which had started (probably in
Japan) in the tenth century and was to last to the beginning of the
twentieth, even enjoying a short period of intellectual respectability
before it expired.

In the course of the tenth century, Japan suffered six epidemics of
smallpox. Buddhist prayers, offerings to Shinto shrines, commutation
of taxes and the granting of amnesties were all tried but none successfully. A book was published which recommended the hanging of red
cloth in the sickrooms of those suffering with smallpox. By the twelfth
century the use of red objects to treat smallpox was being rationalised
by Averroes, a Spanish Muslim philosopher and physician, who
argued that because red objects had warming properties they would
help expel undesirable humours through the skin. By the thirteenth
century Gilbert, an English physician who had studied and lived in
France, was approving the practice of old countrywomen who added
purple or red ingredients to the drinks of those suffering from smallpox. A century later physicians in Montpellier and Piedmont were
using red bed coverings, King Charles V of France was given a red
shirt and stockings, and John of Gaddesden was wrapping the English
King's son with yards of red cloth, and encouraging him 'to suck the
juice of a red pomegranate and to gargle his throat with red mulberry
wine'.[28] (John boasted of a cure but Cyril Dixon, in his classical book
on smallpox, suggests that the prince had probably been suffering from
chickenpox.[29])

In the fifteenth century a Portuguese physician, Valescus de Taran-
ta, recommended wrapping patients in purple or red woollen cloth; in
the sixteenth century, Queen Elizabeth was wrapped in a red blanket;
by the seventeenth the practice had spread to Switzerland, and by the
eighteenth it had reached Austria and Sweden. The British ambassador at Constantinople wrote to the Royal Society in 1755, quoting a
Georgian physician who said his people believed that an angel presided
over smallpox, and that when they inoculated, 'to attract the Angel's
good-will more effectually, they hang the patient's bed with red cloth
or fluff, as a colour most agreeable to him'.[30] In the great eighteenth-
century Chinese novel *The Story of the Stone*, when the infant daughter
of one of the principal characters gets smallpox, 'a length of dark-red
cloth had to be procured and made up into a dress for the child by the
nurses, maids and female relations most closely associated with it'.[31]

By the nineteenth century the red treatment had reached Russia and
Denmark. But the most surprising part of the saga is that in the 1890s
in Denmark, Niels Finsen, who was later to win the Nobel Prize for
showing that ultra-violet light could be used successfully to treat the
tuberculous skin infection *lupus vulgaris*, claimed that if patients with
smallpox are placed

> *in rooms from which the chemical [i.e. short wavelength] rays in the*
> *solar spectrum are excluded by interposing red glass or thick red cloth,*
> *... the vesicles as a rule do not enter upon the stage of suppuration, and*
> *... the patients get well with no scars at all, or at most with extremely*
> *slight scarring.*[32]

For some years these claims seemed to be substantiated by a number
of studies, mainly in Scandinavia, and it was tempting to assume that
exclusion of short-wave light provided a rational basis for the traditional red treatment (although it was doubtful how much protection
traditional procedures would have given). By 1903, though, better
designed studies with more carefully selected patients made Finsen's
claim unconvincing, and the method faded out.[33]

Following the red trail has led us way beyond the proper bounds of this chapter, and we must return briefly to medieval times. Crusaders coming back to Europe in the twelfth century brought smallpox with them. In England, the early presence of smallpox is indicated by two manuscripts in the British Museum.[34] The first, from the end of the tenth century, has a prayer in both Latin and Anglo-Saxon which includes a phrase meaning 'shield me against the loathsome pocks' – in the original Anglo-Saxon: 'geskyldath me vid de lathan Poccas'. The second, from the eleventh century, has a Latin prayer intended for the consecration of protective amulets made by nuns, invoking the help of St Nicaise and including the Latin equivalent of '… may the work of these virgins ward off the smallpox.' There is also a story that Aelfthryth, a daughter of Alfred the Great, survived an attack of smallpox in 907, but as she had by then been long married to the Count of Flanders it is unlikely that she was in England. By the end of the twelfth century, smallpox had spread widely in Europe. In the thirteenth century a Danish ship carried it to Iceland, where it is said to have killed about 20,000 people.[35] Two hundred years later it hit Greenland and Sweden. There is no definite evidence of smallpox in Russia until the seventeenth century, but long before that, as European horizons widened, Europe began to be a source rather than a recipient of smallpox. Just how serious a source will be shown in the next chapter.

There is, though, one other event that belongs to this chapter, and that is said to account for the term smallpox. 'Pox' is a variant spelling of 'pocks', from the old English 'poc' meaning pustule or ulcer; but why 'small'? In 1495, Charles VIII of France, that 'young and licentious hunchback of doubtful sanity',[36] succeeded in invading Italy and getting himself crowned King of Naples without a fight. But success was short-lived, and the same year saw him and his multinational army retreating across the Alps, defeated and with little to show for their efforts. One thing his troops did bring back was syphilis – possibly introduced to Europe by Columbus' sailors returning from America, though that hypothesis is still controversial. The French noted the resemblance between the rashes associated with the two diseases, and distinguished syphilis as '*la grosse vérole*' and smallpox as '*la petite vérole*'. Hence, in English, the great pox and the smallpox.

4

Smallpox in the age of discoveries:
1500–1700

John Simon, Britain's first Medical Officer of Health – and a close
friend of Ruskin and the Pre-Raphaelites – wrote a classic and
dramatic report to Parliament in 1857. It was about the history and
practice of vaccination, and the introduction described the ravages of
smallpox as it had spread from Europe around the world:

> To remote or insular populations, having infrequent and difficult inter-
> course with the busier masses of mankind, such an infection would come
> seldom; but, having come, it would find, perhaps, the entire generation
> prone to receive it ... Thus it was that in 1518, following European
> adventure to the Western world, it concurred with fire and sword and
> famine and blood-hounds to complete the depopulation of St Domingo;
> thus, that soon afterwards, in Mexico, it even surpassed the cruelties of
> conquest, suddenly smiting down 3½ millions of population and leaving
> none to bury them; thus, that in Brazil, in the year 1563, it extirpated
> whole races of men; thus, that about the same period, in the single prov-
> ince of Quito (according to De la Condamine) it destroyed upwards
> of 100,000 Indians. And thus, too, it has been in later days that Siberia
> and Kamschatka have been ravaged; thus, that again and again, till very
> recent times, the same dreadful pestilence has depopulated Greenland
> and Iceland.[1]

Estimates of numbers, both of populations and of deaths, vary wildly.
According to the 1857 report, when Columbus first arrived at San

Domingo (now Haiti and the Dominican Republic), there was a friendly population of a million Indians; the *Encyclopaedia Britannica* (1911 edition) puts it at 2 million, and adds coldly: 'they were, how-ever, soon exterminated'. It is not clear whether smallpox was brought there by West African slaves or by the Spanish themselves, but there were certainly outbreaks from 1507 onwards. A third of the popu-lation died, according to some; only about 1,000 survived according to others.[2] What is certain is that smallpox was carried from there to Puerto Rico (where it is said to have killed over half the native popula-tion) and to Cuba. It was from Cuba that Hernando Cortéz sailed to Mexico, followed soon after by Panfilo de Narváez, with an African slave suffering from smallpox on board.

There are various vague theories about how or when smallpox had reached West Africa, a land with no early written records. Perhaps it had come with the Arabs to North Africa, and then followed the traders across the Sahara; or traders might have brought it from Egypt. By the time the slave ships arrived, spreading smallpox to America and further round Africa itself, it was already endemic in parts of West Africa, and the Yoruba (in modern Nigeria) had a cult of worship of a smallpox god.

Smallpox was endemic in Spain too, for it had been there since the arrival of the Moors, and most of the Spaniards arriving in the New World in the early sixteenth century were immune. So not only were the Mexicans struck down themselves, but to add to their terror they saw that the gods seemed to protect the invaders. The Spaniards saw it that way too – 'When the Christians were exhausted from war', one wrote, 'God saw fit to send the Indians smallpox, and there was a great pestilence in the city.'[3] As a Spanish friar wrote, in 1541, the Indians

> *died in heaps, like bedbugs. Many others died of starvation, because as they were all taken sick at once, they could not care for each other, nor was there anyone to give them bread or anything else. In many places it happened that everyone in a house died, and, as it was impossible to bury the great number of dead they pulled down the houses over them*

5. *A Yoruba smallpox god from Nigeria*

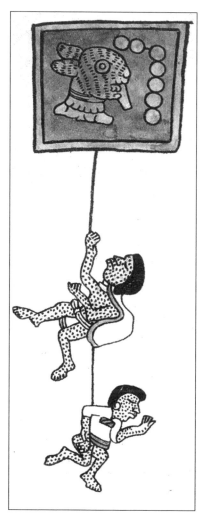

6. *Boys with smallpox in Mexico, 1538*

in order to check the stench that rose from the dead bodies so that their
homes became their tombs.[4]

The estimated original Mexican population of 15 to 30 million shrank
to about 3 million by 1568, and to about 1.6 million by 1620; smallpox
was the principal culprit.[5] Francisco Hernández Arana, grandson of
King Hunyg of Yucatán, wrote:

> *Great was the stench of the dead. After our fathers and grandfathers*
> *succumbed, half the people fled to the fields. The dogs and vultures*
> *devoured the bodies. The mortality was terrible. Your grandfathers died*
> *and with them the sons of kings and their brothers and kings' men. So it*
> *was that we became orphans, oh my sons. So we became when we were*
> *young. All of us were thus. We were born to die.*[6]

Among those who died was the Aztec emperor Cuitláhuac and at
least two other chiefs. Without the help of smallpox, even horses and
guns could not have enabled Cortéz, with his army of fewer than 900
men, to defeat the Aztecs and conquer Mexico.

Epidemics spread from Mexico to Peru before the arrival of
the Spanish conqueror Francisco Pizarro in 1527, weakening the
state, provoking a civil war and leaving the Incas demoralised and
vulnerable. Reliable figures are again elusive, but we are told that
thirty years later the Inca population had been reduced at least
by a half, and possibly by three-quarters.[7] The terrible saga of
destruction of the Indian civilisations rolled on through the continent, to Guatemala, Ecuador, Colombia, Chile. The Portuguese
took smallpox to Venezuela, the French and Portuguese took it to
Brazil. It came a little later to Argentina, when the arrival of slave
ships in Buenos Aires in 1603 set off the first of seven epidemics
there that century. It was a century that saw continuing smallpox
epidemics throughout the Latin American colonies. Jesuit missionaries made gallant efforts, in the most appalling conditions, to
save the bodies as well as the souls of the Indians, but the sad

fact is that by assembling the Indians at missions they probably made matters worse.

Shortage of labour led to the large-scale importing of Africans to work on the plantations and in the mines. Some of these Africans arrived immune; others, like that first slave who had crossed to Mexico with Narváez, added to the problem by bringing smallpox with them. The first mention of slaves brought to work on the Spanish estates is in San Domingo as early as 1503, but the smallpox depopulation was a major factor leading to the growth of the traffic through the following three centuries, with its endless suffering and its political and economic complications.

In North America the story was different, though no less tragic. Here, settlers came in groups, a hundred years later than Cortéz' invasion of Mexico, to a land where the Indians were just as vulnerable, but in more scattered communities. 'Princess' Pocahontas, who had befriend-ed the Virginian settlers, had died of smallpox in 1616 when she was only twenty-one, but that was on a visit to England just before small-pox arrived in North America. The Pilgrim Fathers, landing in 1620, found the area round Plymouth nearly deserted; an epidemic a year or two earlier, probably smallpox coming from the French settlement in Nova Scotia, had killed almost nine-tenths of the Indians along the Massachusetts coast.[8] Smallpox spread, and North American Indians, like those in the South, died in their thousands; epidemics hit Massa-chusetts' Indians in 1633, Narraganset county and Connecticut Indi-ans the following year, and then on westward and to the Great Lakes and the St Lawrence. Numbers here are even more uncertain than in the South, but French Jesuits left records of tribes weakened and fami-lies and villages destroyed. The Iroquois, the Mohawks, the Mohegans, the Algonquins – all these evocative names appear in the stories of suffering. Like the Spaniards, the new settlers found such epidemics providential. John Winthrop, who had emigrated in 1630 and had become governor of the new colony of Massachusetts, wrote four years later that, 'the natives, they are neere all dead of the small Poxe, so as the Lord hathe cleared our title to what we possess';[9] and the Puritan preacher Increase Mather had the same thought when he looked back at the early settlers from later in the century – 'the Indians began to be quarrelsome touching the Bounds of the land which they had sold to the English; but God ended the controversy by sending the smallpox amongst the Indians'.[10]

Travel, by traders, settlers, soldiers, or by the Indians themselves, helped to spread recurrent epidemics throughout the seventeenth century in Canada and in New England. French Jesuits, who first went to Canada in 1625, were, like the Spanish Jesuits in Latin America, the unwitting source of terrible disasters as they went from village to village, conscientiously baptising and sick-visiting. Puzzled at the hostility they met, they admitted that the Indians 'observed with some sort of reason that since our arrival in these lands those who had been the nearest to us had happened to be the most ruined by the disease, and that whole villages of those who had received us now appeared utterly exterminated'.[11] As one group told them, 'This disease has not been engendered here: it comes from without; never have we seen demons so cruel.'[12] Matters were made worse by deliberate infections by white colonists – 'numerous instances', one book on the American Indian claims, 'of the French, the Spanish, the English, and later on the American, using smallpox as an ignoble means to an end'.[13] And the Indians had no tradition of managing infections by isolation; the Jesuits reported in 1640 that 'the Hurons, no matter what plague or contagion they may have, live in the midst of their sick in the same indifference and community of things'. It was not only the Hurons who lived that way – and tribe after tribe suffered.

The English and the French had less experience of smallpox than the Spaniards, and so had less immunity, but they were better than the Indians at recognising the need for isolation when epidemics struck. Trouble came early, when twenty settlers from the *Mayflower*, including their only physician, died in the 1633 epidemic; some, including thirty from William Penn's party in 1682, died of smallpox on their way across the Atlantic. The populations were still too small for the infection to be maintained endemically, which meant that many adults as well as children were vulnerable. At first they sought protection through prayer and fasting, but these early efforts were soon supplemented by the more practical and long-established ideas of isolation and quarantine. Many people in Africa, Asia and Europe had, far earlier, tried to avoid leprosy and plague by isolating the victims in separate huts or villages. The fourteenth-century Venetians started isolating travellers and ships infected with plague for a forty-day period, and the idea of quarantine (*quaranta* being Italian for forty) was taken up by European colonists for ships suspected of carrying smallpox. Quarantine measures were first set up in Boston in 1647 for ships arriving from the West Indies. In small settlements and scattered populations the idea of containing the infection was practical; in the eighteenth century, as the towns grew, the outbreaks became more difficult to control.

The infection the settlers had brought to the Indians came back to hit the settlers themselves; in 1662 the town of East Hampton, Long Island, ordered that no Indian should come into town 'upon a penalty of 5s or to be whipped, until they be free of the small poxe ... and if any English or Indian servant shall go to their wigwams they shall suffer the same punishment'.[14] In 'King William's War', when the English fought the French, each with their Indian allies, both sides were heavily hit by smallpox. The English, allied with the Mohegans and Iroquois, were forced to abandon a planned attack on Quebec in 1690; the Jesuit report said that the British had sent both an army and a navy against Quebec, 'smallpox stopped the first completely and also scattered the second'.[15]

Unlike the explorers of the New World, the sixteenth-century explorers who travelled East, even though they often took smallpox with them, did not usually cause such devastation because they came to lands where smallpox was already endemic. Very few Europeans reached China or Japan, but the Portuguese in the East Indies, the Spaniards in the

Philippines, the Portuguese, Dutch and British in India and Ceylon, all found that smallpox had got there before them. In isolated communities though, the disease could die out for long enough between epidemics to leave most of the population vulnerable – in 1591 a Jesuit in Manila reported that 'one third of the population was in bed and there was not left any person who was not attacked by it and many died, especially among adults and the aged'.[16] A King of Burma had died of smallpox in the fourteenth century, a King of Siam, and the King and Queen of Ceylon together with all their sons in the sixteenth. In India there are many tales of outbreaks disastrously attacking princes and generals – five of the sixteen maharajahs of Tripura between the fifteenth and eighteenth centuries died of smallpox.[17]

In Goa, where smallpox must have been endemic and the adults therefore largely immune, 8,000 children died in the epidemic brought by the Portuguese in 1545. In the seventeenth century, a European visitor to the east coast wrote that 'smallpox invades the youth, as in all India'.[18] The British settled, with varying degrees of sympathy, into accepting smallpox as part of life in the East. Mansell Smith, running a factory at Calicut on the west coast of India in 1675, wrote of an epidemic that had attacked most of the house servants; three had died, 'so', he added, 'we are very hard pressed to get our victuals dressed'.[19]

Smallpox was at that time also very much part of British life at home. There is a theory that, having been relatively mild in Europe until the seventeenth century, it had become a more severe disease – though it had not seemed mild to Queen Elizabeth, or to her suitor, Alençon, who had been left horribly scarred, or to the son of Francis I of France who was blinded in one eye, or to the many who died in sixteenth-century epidemics in the towns of Italy and Spain. But smallpox certainly appears often and alarmingly in the letters and diaries of seventeenth-century England. 'The disease itself in its own nature is now become ordinarily very mortal, especially to those of your age', the royalist lawyer Matthew Hale wrote to his grandson.

Look upon even the last year's general bill of mortality, you will find near two thousand dead of that disease the last year; and, had God not been very merciful to you, you might have been one of that number ... Your sickness was desperate, in so much that your symptoms and the violence of your distemper were without example; and you were in the very next degree to absolute rottenness, putrefaction, and death itself.[20]

There is a seventeenth-century monument in Great Barrington, Oxfordshire, to the Bray family, who 'had five sons and two daughters, and lost six of them from smallpox'. And smallpox had become notably more widespread, as well as more severe. A 'Wanted' notice in the *London Gazette* describing a counterfeiter, mentioned as a distinguishing feature that he was 'without pock holes'.[21] Pepys remarked in 1668 that 'it hardly ever was remembered for such a season for the smallpox as these two months have been';[22] the King's mistress, the Duchess of Richmond (model for the figure of Britannia on the English penny), was among those who suffered, and it 'would make a man weep to see what she was then and what she is like to be, by people's discourse, now'.[23] As that other great diarist, John Evelyn, wrote, 'The smallpox increas'd exceedingly, and was very mortal.'[24]

Poor Evelyn, like many others, knew only too much about the disease. He had been miserably ill from it himself nearly fifty years earlier, and in March 1685 his 'deare, sweete, and desireable' daughter Mary had died of it at the age of nineteen. In July that same year smallpox took her friend Mr Hussey, who had hoped to marry her; and only a month later he was followed by Evelyn's recently married daughter Elizabeth. 'Thus in less than six monthes were we deprived of two children for our unworthinesse and causes best known to God.'

The royal House of Stuart could tell much the same story. Macaulay described how, when 'the most terrible of all the ministers of death' struck Queen Mary II, she

gave orders that every lady of her bedchamber, every maid of honour, nay, every menial servant, who had not had the smallpox, should instantly

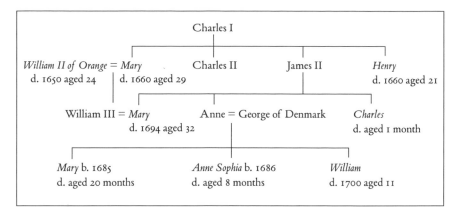

```
                            Charles I
        ┌──────────────────────┼──────────────┬──────────────────┐
William II of Orange = Mary      Charles II      James II        Henry
  d. 1650 aged 24  │  d. 1660 aged 29              │        d. 1660 aged 21
                   │                      ┌────────┴──────┐
                   └──────┐               │               │
       William III = Mary           Anne = George of Denmark    Charles
              d. 1694 aged 32              │              d. aged 1 month
      ┌──────────────────┬─────────────────┴──────────┐
   Mary b. 1685      Anne Sophia b. 1686         William
   d. aged 20 months   d. aged 8 months        d. 1700 aged 11
```

7. Stuart family tree (simplified); all those in italic type died from smallpox

leave Kensington House. She locked herself up during a short time in her closet, burned some papers, arranged others, and then calmly awaited her fate.

Charles II and his sister Henrietta had both earlier survived an attack, but his brother the Duke of Gloucester and his sister Mary (mother of William III, whose father had died of smallpox ten years earlier) had both died in 1660. James II's baby son Charles died from smallpox. Anne, who had lost two baby girls from the disease, suffered the final blow to the Stuart dynasty two years before she became Queen, with the death of young Prince William.[25]

Where statistics exist, as in London and in Geneva, smallpox appeared increasingly to replace plague as the principal cause of death. The figure for annual smallpox deaths in London was seventy-two in 1633, but it was over 1,000 for the first time in the epidemic year 1634, and in five other years before plague had its final explosion in 1665.[26]

While the explorers were spreading smallpox with such devasta-tion round the world, increased mobility within Europe was help-ing its spread in a less dramatic way. There was plenty of mobility among armies – with the French religious wars of the sixteenth century, the German armies of the Thirty Years War (1618–48), or the Russian conquest of Siberia. And in England at this time the growth of population – it roughly doubled between 1520 and 1680 – was encouraging the development of weaving as a cottage industry, and was bringing migrants to the towns in search of work; London grew tenfold, from less than 60,000 in 1520 to 575,000 in 1700,[27] and there were by then seven other English towns with a population of over 10,000. The trade engendered by all these towns and new industries made matters worse because merchants travelling around the country carried infections wherever they went.

———

There was of course no sign of a cure for smallpox, but there was some progress in the understanding of infection. A theory that contagion could come in three ways had first appeared in fifteenth-century Salerno: 'children should avoid touching the contagion of the disease – viz. (1) the sick person, (2) the breath of the sick, and (3) the clothes, coverings, garments, and such clean bodies as he may have infected'.[28] Girolamo Fracastoro, a hundred years later, carried these ideas further, believing that smallpox and measles each had its own specific 'seeds' (seminaria) of infection which could, unlike a mere poison, multiply. Oddly, the great seventeenth-century physician Thomas Sydenham, often clear and analytical in his approach, confused the infection issue with obscure thoughts about 'hidden and inexplicable changes within the bowels of the earth' that affected what he called the 'epidemic constitution' of the atmosphere. 'As to what may be the essence of smallpox', he wrote, 'I am, for my own part, free to confess that I am wholly ignorant.'[29]

Sydenham, who studied medicine after a hazardous time fighting on the Parliamentary side in the Civil War, was a thoughtful and observant physician. He is remembered for his descriptions of gout (which he suffered from himself) and St Vitus' dance (which is still referred to as 'Sydenham's chorea'), but his useful contributions to smallpox are that he distinguished between the less dangerous discrete form, and the more dangerous confluent form; and he gave the welcome advice that the heat regime, which had been customary since the time of Rhazes, should give way to a 'cooling method', with fresh air and light coverings. Noting that the rich, who could afford treatment, died of their smallpox more readily than the poor, he thought there was a danger in too much interference. He was bold and original enough to advise that as soon as the smallpox was recognised as discrete, it should be left untreated – not an easy line to take, for as James Moore wrote 150 years later, 'to do nothing is frequently the last improvement made by physicians; and one which patients very rarely can be induced to acquiesce in'.[30] Even Sydenham did not abandon bleeding altogether, but he used it less vigorously, and usually only in the early stages of the disease. This part of his thinking dated back to Rhazes but, unlike Rhazes, Sydenham kept to the cooling regime throughout the disease – a practice new in Europe, though it had long been recommended in India, where the smallpox goddess Sitala, with her water and her fan, was worshipped as 'the quencher of the fierce heat of pustules'.[31]

Sydenham was very ready with vomits and purges, and even sometimes bleeding, in confluent cases. His pupil Thomas Dover caught 'anomalous smallpox' (part discrete and part confluent) and has described his treatment:

> *Whilst I lived in Dr Sydenham's house, I had myself the Small Pox, and fell ill on the Twelfth Day. In the beginning I lost twenty two Ounces of Blood [i.e. he was bled]. He gave me a Vomit, but I find by Experience Purging much better. I went abroad, by his Direction, till I was blind, and then took to my Bed. I had no Fire allowed in my Room, my Windows were constantly open, my BedClothes were ordered to be laid no higher than my Waste. He made me take twelve Bottles of Small Beer, acidulated with Spirit of Vitriol, every twenty Four Hours. I had of this Anomalous Kind to a very great Degree, yet never lost my Senses one Moment.*[32]

(Dover was more notable for his eventful life than for his medical skill. A relentless believer in bleeding, known as the 'quicksilver doctor' for his faith in mercury, he became famous for a prescription containing opium and ipecacuanha, known as 'Dover's powder', and he denounced the College of Physicians as a 'clan of prejudiced gentlemen'. But he spent three years of his life as second in command of a privateering voyage round the world, rescuing the shipwrecked sailor Alexander Selkirk (whose story inspired Defoe to write *Robinson Crusoe*), plundering a city in Peru, and capturing a Spanish prize ship.)

A personal quarrel, with arguments over the relative merits of vomits and purges as treatment for smallpox, on one occasion even led to blows. In 1719 Richard Mead, eminent physician, Vice-President of the Royal Society, and a purging man, attacked John Woodward, eminent geologist, Professor of Physic at Gresham College, and a vomit man – the physician John Freind called him 'Dr Emeticus'. Mead considered Woodward 'a man equally ill-bred, vain and ill-natured'[33] – as indeed he had shown himself to be on an earlier occasion when he had insulted Sir Hans Sloane at the Royal Society council and, refusing to apologise, had been expelled from the council. (Mead, on the other hand, must have had a more amenable nature, as he remained a life-long friend of the notoriously difficult Richard Bentley, Master of Trinity College Cambridge.) Mead and Woodward fought with swords in the quadrangle of Gresham College until Woodward slipped and fell, and bystanders intervened to keep the peace.

Smallpox deaths rose to over 3,000 in London in the epidemic year of 1710. Little real help was offered by any of the treatments the doctors advocated, and a good deal of harm was done by some. Luckily, news started to reach London of the possibility of prevention.

5

News from the East

The smallpox, so fatal and so general amongst us,
is here entirely harmless by the invention of engrafting.
*Lady Mary Wortley Montagu, wife of the British ambassador to Turkey, in a letter
to Sarah Chiswell, from Constantinople, 1 April 1717*

Lady Mary Wortley Montagu, famous for promoting inoculation in England, used the gardening term 'engrafting' to describe it. And 'inoculation' is itself in origin a gardening term, literally meaning 'in-eyeing', the insertion of a bud.

You might expect an eighteenth-century British ambassador's wife to be an upper-class pillar of the establishment. Lady Mary was certainly upper class, but she was something of a professional rebel. She eloped with Edward Wortley Montagu when her father, Lord Dorchester, had forbidden the match and had arranged for her to marry the Honourable Clotworthy Skeffington, who had nothing in his favour but his wealth – hardly suitable for someone who was to write that she 'prefer[red] liberty to chains of diamonds'.

'Every branch of knowledge is entertaining and the longest life is too short for the pursuit of it', Lady Mary wrote much later, and she showed admirable curiosity and understanding in her travels in Turkey. The portrait frontispiece to her letters shows her in turban and Turkish dress, and in a letter written to her sister from Adrianople, Lady Mary described her Turkish clothes in detail, beginning with 'a pair

8. Lady Mary Wortley Montague in Turkish dress

of drawers, very full, that reach to my shoes and conceal the legs more modestly than your petticoats'.

Her defiance of convention and her interest in local customs combined to rouse her interest in the Turkish practice of inoculation – the insertion of a small amount of matter from a pustule of a smallpox patient into the skin of a healthy subject. (The matter is sometimes referred to as *variolous matter* and the process of inoculating it into a healthy subject later became known as *variolation*.)

This was the 'engrafting' that Lady Mary was reporting in her letter to her childhood friend Sarah Chiswell. It was a type of folk-medicine, a technique practised not by physicians but by peasant women, which had been used with 'happy success' on 'thousands of subjects'[1] and seems to have been widespread among the Arabs and among the Christians living in Turkey, but not among the Muslim Turks.[2] Why injection of variolous matter into the skin generally caused only a very mild attack of smallpox is still not fully understood, but part of the attraction to Lady Mary was that it was unorthodox medicine, which she hoped to promote on her return by going to war with the medical establishment. 'There is a set of old women', she wrote,

> *who make it their business to perform the operation. Every autumn, in the month of September, when the great heat is abated, people send to one another to know if any of their family has a mind to have the smallpox. They make parties for this purpose, and when they are met (commonly fifteen or sixteen together) the old woman comes with a nutshell full of the matter of the best sort of smallpox and asks what veins you please to have opened. She immediately rips open that you offer to her with a large needle (which gives you no more pain than a common scratch) and puts into the vein as much venom as can lie upon the head of her needle, and after binds up the little wound with a hollow bit of shell, and in this manner opens four or five veins.*

In this way 'matter of the best sort' – that is, from a mild case of small-pox – could be engrafted at a time when the patient was fit to receive it, though it was not quite the jolly event that the letter light-heartedly describes. Nor did the operation involve ripping open a vein; it was the insertion of a small amount of matter from a pustule into a shallow prick in the skin.

Charles Maitland, the physician at the embassy in Constantinople, described the ordeal of Lady Mary's own six-year-old son:

> *She first of all order'd me to find out a fit Subject to take the matter from: and then sent for an old Greek woman, who had practis'd this Way a great many Years: After a good deal of Trouble and Pains, I*

found a proper Subject, and then the good Woman went to work; but so
awkwardly by the shaking of her Hand, and put the Child to so much
Torture with her blunt and rusty Needle, that I pitied his Cries …
and therefore inoculated the other Arm with my own Instrument, and
with so little Pain to him, that he did not in the least complain of it.
The Operation took in both Arms, and succeeded perfectly well. After
the third Day, bright red Spots appear'd in his Face, then disappear'd
… betwixt the Seventh and Eighth Day … the Small Pox came out
fair … the young Gentleman was quickly in a Condition to go Abroad
with safety. He had above an hundred [pocks] in all upon his Body.[3]

(This inoculation not only protected young Edward from smallpox;
he ran away to sea at the age of fifteen and when, tired of disguise, he
made himself known to the ship's captain, he used his inoculation scar
to prove his identity.[4])

There are stories of variations on the inoculation theme wherever in
the world smallpox was endemic, from Wales to China. Why did this
tradition develop, and why did it develop only for smallpox? Perhaps
it started because smallpox was so terrible and seemed so inevitable,
that children were sometimes deliberately exposed to the infection, in
the belief that it was best to catch it when young and from a mild case.
John Evelyn wrote of calling on a former maid of honour to the Queen
where the

eldest son was now sick of the smallpox, but in a likely way to recover,
and others of her children ran about and among the infected, which she
said she let them do on purpose that they might whilst young pass that
fatal disease she fancied they were to undergo one time or other, and that
this would be for the best.[5]

As late as 1784 John Haygarth, the first doctor to treat fever patients in
separate wards, reported that in Chester 'No care was taken to prevent
the spreading; but on the contrary there seemed to be a general wish
that all the children might have it.'[6] (Even in New York in the middle

of the nineteenth century a case was recorded of a mother, not trusting vaccination, deliberately exposing her four children to smallpox; three of them died.[7])

Deliberately exposing children to the disease may have been a small step towards the custom of 'buying the smallpox' — a strange custom made stranger by the unexplained fact that almost identical rituals evolved in places as unconnected as Algiers (Muslim Arabs, unlike the Turks, did practise inoculation), West Africa, Switzerland, Wales and Poland. 'The child to be inoculated', wrote Patrick Russell reporting a tradition in the Syrian town of Aleppo in 1768,

> *carries a few raisins, dates, sugar plumbs, or such like, and shewing them to the child from whom the matter is to be taken, asks how many pocks he will give in exchange. The bargain being made they proceed to the operation. When the parties are too young to speak for themselves, the bargain is made by the mothers.*[8]

And he added that inoculation was general in Baghdad, in Georgia, in Armenia and in Mecca.

In the same year, a report from North Africa described the formalities of the bargain:

> I am come here to buy the small-pox: *the answer is,* buy if you please. *A sum of money is accordingly given, and one, three, or five pustules (for the number must always be an odd one, not exceeding five), extracted whole, and full of matter. These are immediately rubbed upon the skin of the hand, between the thumb and fore-finger. This is sufficient to communicate the infection; and as soon as it begins to take effect, the inoculated patient is put to bed, carefully covered with red blankets; and heating medicines are given him with some honey of roses. He is allowed goat's broth for his nourishment, and for his drink an infusion of some herbs; notwithstanding this treatment, it seldom happens that the small-pox procured in this manner has any bad consequences.*[9]

The Danish anatomist Thomas Bartholin saw the origin of this idea of transference of the disease in the biblical story of the scapegoat[10] – though no one could suggest that the biblical transfer was for the sake of the goat. There was ritual and superstition too in the Greek version of inoculation, where the women made the incisions in the form of a cross, on the forehead, arms and breast, with prayers and the offering of candles.

An alarming account written to the Royal Society in 1723 described the experience of the son of the archdeacon of St David's, in Wales, who:

> *when at School, rubb'd the Skin off his Left Hand, where the Scar is now very visible, with the back Edge of his Penknife, till the Blood began to apear; he apply'd the variolous Matter [taken from a pustule of a patient with smallpox] to that part; which by Degrees growing inflam'd, about a Week afterwards he fell into the Small Pox ... He says also five or six more at least of his Schoolfellows made the like Experiment on themselves, at the same time, with the like Success.*[11]

The usual mildness of inoculated smallpox is often attributed to the abnormal route of entry of the virus (and perhaps also the small dose) for many microbes are less effective when they enter the body by an unusual route. In China, though, where smallpox had long been endemic, the technique was different. There are various stories about the earliest inoculations, but they all put them at around 1000 AD when, in different versions, a nun, or a holy physician, or a Buddhist master, or an ancient immortal is said to have been summoned from southwestern Szechwan to save the son of a great statesman from infection.[12] Descriptions of the method are more consistent. It was done through the nose, either by stuffing up plugs of cotton impregnated with powder from scabs, or, more commonly, by blowing the powder into the patient's nostrils through a silver tube – in the right nostril for a boy, the left for a girl. Here then, the mild response cannot be explained by an abnormal route of entry, so why was it mild? According to the biochemist and sinologist Joseph Needham, scabs were taken only from patients with mild inoculated smallpox, and it was the custom to keep the powder at close to body temperature against the variolator's chest for a month or more. This would have killed most, though not all, of the viruses, and he suggests that the dead virus may have still been able to induce immunity.[13] (Pasteur first succeeded in making a vaccine against rabies by drying the spinal cords of infected rabbits long enough to make extracts from them uninfective yet still able to induce immunity in dogs into which they were injected.[14]) It may also be relevant that unlike inhaled fine droplets, which can spread through the entire airway, an inhaled powder may be largely trapped and removed by the lining of the upper part of the airway.

It was not until the sixteenth century that inoculation was written about in Chinese books on medicine and became more widely practised. Li Shih Chen wrote an encyclopaedia of pharmacology, with 8,000 prescriptions, and a description of inoculation.[15] China was at that time notorious among her Mongol and Manchu neighbours for the danger of infection – officials who had not had smallpox avoided going to court in Peking. And when the Manchus overthrew the Ming dynasty, their Emperor Fu-lin caught smallpox and died in 1661. Fu-lin had eight sons, and it was the eight-year-old third son, who had already had smallpox, who was chosen to succeed him – by a great piece of luck Emperor K'ang Hsi, chosen in this eccentric way, turned out to be one of China's finest rulers. Not surprisingly, he was concerned with the problem of smallpox:

> *The method of inoculation having been brought to light during my reign,*
> *I had it used upon you, my sons and daughters, and my descendants, and*
> *you all passed through the smallpox in the happiest possible manner ...*
> *In the beginning, when I had it tested on one or two people, some old*
> *women taxed me with extravagance, and spoke very strongly against*
> *inoculation. The courage which I summoned up to insist on its practice*

9. Artist's impression of inoculation by insufflation in China

*has saved the lives and health of millions of men. This is an extremely
important thing, of which I am very proud.*[16]

As Needham has pointed out, Emperor K'ang Hsi's experiments fore-
shadowed similar experiments forty years later in London.

The Japanese did not learn from the Chinese about inoculation
until the middle of the eighteenth century. But in the thirteenth century smallpox had been described there as a children's disease, which
implies that it was by then endemic. With their crowded population
(about 25 million in 1709), the Japanese suffered badly, and they tried
to tackle the problem by abandoning infected households or villages,

and with isolation hospitals. Smallpox killed the promising young Emperor Gokomyo in 1654, the Emperor Higashiyama in 1709, and two of the most powerful seventeenth-century shoguns. By that time the Japanese had learned to distinguish between smallpox, measles and chickenpox, while syphilis (the great pox) became known as 'the Portuguese disease'. In 1796 a chair for the study of smallpox was established at Tokyo's Medical Academy.

East India Company officials wrote from India in the eighteenth century with accounts of inoculations. One, from Bengal, describes a practice similar to that in Turkey, which he thought had been used for about 150 years; some say the tradition went as far back as the eleventh century:

> *Their method of performing this operation is by taking a little of the pus (when the smallpox are come to a maturity and are of a good kind) and dipping in these the point of a pretty large sharp needle. Therewith make several punctures in the hollow under the deltoid muscle [in the upper arm] or sometimes in the forehead, after which they cover the part with a little paste made of boiled rice.*[17]

This practice, Holwell says:

> *... is performed in Indostan by a particular tribe of Bramins, who ... dividing themselves into small parties of three or four each, plan their travelling circuits in such wise as to arrive at the place of their respective destination some weeks before the usual return of the disease.*[18]

And Matthew Maty, physician and librarian at the British Museum, reported to the Royal Society in 1768 on two other systems of inoculation in Bengal – either by pricking between two fingers, or,

> *as this way of managing the operation is very painful, a more easy one has been invented for people of quality and substance. A little of the matter is mixed with sugar and swallowed by the child in any sweet*

and pleasant liquid. The same effect is produced, but the first method is thought to be best.

It was probably from India that knowledge of inoculation spread to the Balkans and to Turkey.

———

Late in the seventeenth century the first hint of inoculation appeared in medical writing in the West, in Bartholin's essay 'On the Transplantation of Disease'. But the Royal Society, which had been founded by Charles II, had become the most important centre for scientific knowledge, and it was to the Society that the first reports of the Chinese practice came in 1700.[19] In 1713 the Italian physician Timoni sent the Society his account of inoculation in Constantinople, and three years later this was followed by an account written by the Greek physician Pylarini. In Boston, the Revd Cotton Mather, Increase Mather's fiery son, saw the paper by Timoni, and realised that it described something he had already learned about from his slave Onesimus, who had recently arrived, with an inoculation scar on his arm, from southern Tripoli. Mather sent the story of Onesimus to the Royal Society in an essay on 'Curiosities of the Smallpox' (1716), saying that it was 'the same that afterwards I found related unto you by your Timonius', and asking, 'How does it come to pass that no more is done to bring this operation into experiment and into Fashion?' 'For my own part,' he added, 'if I should live to see the smallpox again enter into our city, I would immediately procure a consult of our physicians, to introduce a practise which may be of so very happy a tendency.'[20]

So when Lady Mary Wortley Montagu wrote to Sarah Chiswell in 1717, two weeks after her arrival in Turkey, she was describing something strange, but not totally unheard-of in England. And when she had her own little son inoculated in Constantinople the following year, she was not the first European or even the first British ambassador's wife to do so – inoculation had been successfully performed on the two sons of the previous ambassador. But Lady Mary had written in that letter that she was 'patriot enough to take pains to bring this useful invention into fashion in England'. By her example and by her energetic propaganda in aristocratic circles, that is just what she did.

Lady Mary had every reason to feel strongly about smallpox. Her brother had died from it in 1713 at the age of twenty, and she herself, once well known for her beauty, had an attack two years later which left her badly scarred and without eyelashes. So when she was back in England in 1721 she persuaded Charles Maitland, who had also returned, to inoculate her three-year-old daughter – the first time this had been done in England by a member of the medical profession.

And it was done with maximum publicity. Maitland, worried about the responsibility, insisted on having two physicians as witnesses 'not only to consult the Health and Safety of the Child, but likewise to be Eye-Witnesses of the Practice and contribute to the Credit and Reputation of it'. More physicians (one of whom then got Maitland to inoculate his own son) and 'other persons of distinction' came to inspect the child's pocks. After a slight fever young Mary recovered well, and lived to marry Lord Bute (later George III's prime minister) and to bear eleven children. The inoculation was done, as Sir Hans Sloane wrote, 'by making a very slight shallow incision in the skin of the arms about an inch long', taking great care 'not to go thro' the skin'.[21]

One person of distinction who was interested was Caroline of Ansbach, the highly intelligent Princess of Wales. She herself had had the disease shortly after her marriage and, according to one account, her daughter Anne was just then severely ill with suspected smallpox. 'To secure her other children, and for the common good', Sloane wrote, she 'begged the lives of six condemned criminals, who had not had the small-pox, in order to try the experiment of inoculation upon them'. Anne may or may not have been ill just then, and it is unclear whether the idea of the experiment came from the princess or from physicians, but the experiment on the Newgate prisoners certainly took place.[22] After advice from Edward Terry who 'had practised physic in Turkey [and] had seen the practice there by the Greeks encouraged by their patriarchs; and that not one in eight hundred had died in that opera- tion', Sloane persuaded Maitland to go ahead. There were plenty of witnesses – Sloane himself and twenty-six other medical men – to see that the prisoners had a proper reaction; all did, except one who had previously had smallpox, and all recovered. Richard Mead tried the Chinese method on another prisoner, which 'gave great uneasiness to the poor woman'. All were released. To make sure that the treatment had been effective, Sloane and a colleague paid for one of the prisoners to go to Hertford 'where the disease in the natural way was epidemical and very mortal, and where this person nursed and lay in bed with one, who had it, without receiving any new infection'. Sir John Vanburgh, who had his own son inoculated the following year, wrote to Lord Carlisle:

> *I have seen the Physitians, and askt them how inoculating has really*
> *succeeded, and they assure me, not one Single Person has miscarried,*
> *nor that they find any Sort of ground to fear that those who go thorough*
> *the Small Pox that way, will have them again.*[23]

The idea of trying out inoculation on prisoners carries worrying echoes of Mengele and Nazi experiments, and it might not help to know that it was done with the agreement of the lawyers and the King. But there was every reason to believe that the Newgate prisoners would only gain from the experiment – they would emerge free and permanently immune to smallpox. A London paper even mocked that 'after this Experiment is fully made and approv'd of, any Person that expects to be hang'd may make use of it, if they please'.[24]

The princess promoted a further experiment before she felt ready to let the doctors loose on her own children. She 'procured half a dozen of the charity-children belonging to St James's parish, who were inoc- ulated, and all of them, except one (who had had the smallpox before, tho' she pretended not, for the sake of the reward) went thro' it with the symptoms of a favourable kind of that distemper'.[25] Now the royal inoculations could go ahead.

6

Kicking against the pricks

The inoculation of Princess Caroline's children was carried out successfully by the King's surgeon, Claude Amyand, after consultation with Sloane, and under Maitland's supervision. The publicity has been compared to Queen Elizabeth's 1957 announcement of the polio inoculation of Prince Charles and Princess Anne, which 'was expected to do more than any medical proof in convincing doubtful parents that their youngsters should be given the vaccine'.[1] Sadly, in 1721, confidence was checked almost immediately by the death of two-year-old William Spencer, the Earl of Sunderland's son, eighteen days after his inoculation – in fact he had nearly recovered from the inoculation, and died from possibly unrelated fits. Then came the death of Lord Bathurst's nineteen-year-old footman, blamed on his inoculation, but more likely (again in view of the timing) the result of smallpox caught from the Bathurst children, who were infectious following their inoculation. In spite of these well-publicised tragedies, many aristocratic families did risk following Princess Caroline's example.

It was known that smallpox could normally be caught only once, and that the mild form produced by inoculation gave similar protection, though no one at that time knew why. Physicians were working in the dark, building up their knowledge as they went. When 'persons of distinction' went to inspect Miss Wortley Montagu's pocks, they presumably were unaware that inoculated smallpox could be as infectious as smallpox caught in the natural way. It is odd that this hazard is not mentioned in any of the accounts of inoculation as folk medicine – it must have been known in Turkey, for Lady Mary, when writing to her husband about her son's operation, added, 'I cannot engraft the girl; her nurse has not yet had the small pox.'[2] Maitland, in his 'Account of Inoculating the Smallpox' (1722), told how he had inoculated children in one household in Hertford, who had then infected six servants; and how in another household the inoculated children infected their infant brother. North American physicians in Boston were already worried about infection from inoculated smallpox, but at this time Maitland and his colleagues in England, rather than being concerned about the risk, seemed pleased with this proof that inoculation produced true smallpox. He did comment later though, that 'they never expected them to be catching, nor indeed did I'.[3]

'I have thought [inoculation] an experiment of great consequence to mankind, and therefore have forwarded it all I could,' Sloane wrote to his doctor friend Richard Richardson.[4] There is debate about the relative importance of Sloane and Lady Mary Wortley Montagu in promoting inoculation. They were not friends, for Sloane was the most prominent representative of Lady Mary's scorned medical establish, ment. He had visited her, uselessly, when she had her attack of small, pox:

> In tears surrounded by my friends I lay,
> Mask'd o'er and trembling at the sight of day;
> Mirmillio [Sloane] came my fortune to deplore,
> (A golden headed cane well carv'd he bore)
> Cordials, he cried, my spirits must restore:
> Beauty is fled, and spirit is no more!

But what else could he have done? At least she seems to have been spared his prescriptions of crabs' eyes, oil of scorpions or 'fifty live millipedes in a glass of water twice a day'.

It is easy to make fun of such remedies, but Sloane, who had been a pupil of Thomas Sydenham, was a distinguished physician by the standards of his day, anxious 'to purge the town of bad medicines', to

10. *Bust of Hans Sloane*

11. *Portrait of Cotton Mather*

introduce new herbs he had learnt about in his travels and to encourage the use of quinine. (Having come across cocoa in Jamaica, he thought of tackling its bitterness by mixing it with milk – creating a drink that was sold by apothecaries 'for its lightness on the stomach and its great use in all consumptive cases'; a hundred years after his death Cadbury's was to use his recipe and his name, advertising 'Sir Hans Sloane's milk chocolate'.) By 1721 he had both a fashionable and a charitable practice – he returned his salary from Christ's Hospital, saying 'I shall never have it said of me that I enriched myself by giving health to the poor.' And like many of his great contemporaries in the early days of scientific discovery and the Royal Society, Sloane had interests which spread over all branches of natural history, particularly, in his case, to botany and to building up collections of 'natural and artificial curiosities'. He collected obsessively, and bought up other people's collections. After his death, £95,000 was raised by lottery to buy his collection for the nation, together with the Harleian manuscripts (collected by the statesman Robert Harley, first Lord Oxford, and his son Edward) and Montague House as a home for them all – the beginning of the British Museum.

This is a long way from smallpox, but it shows something of the man who had bought the manor of Chelsea in 1713 (complete with the Physic Garden which he handed over to the Apothecaries' Society), had become President of the College of Physicians in 1719, and was to succeed Newton as President of the Royal Society in 1727. While Lady Mary heard about inoculation from seeing the traditional customs in Turkey, Sloane learnt about it from the communications sent to the Royal Society. Whichever of them was more responsible for the experiments and the royal inoculations, the important point is that the way was now open for the practice, and the arguments, to flourish. As Sloane wrote in 1722:

> *Many physicians, surgeons, apothecaries and divines seem to oppose it with greater warmth than in my opinion is consistent with sound reason, or the good of mankind. What turn it will take time will discover. The*

> Cortex Peruviana *[Peruvian bark, the source of quinine]* met the
> same usage at first entrance into Europe; but it hath been in England
> received with a general applause of late years, though it hath still more
> enemies than it deserves.[5]

England was not quite the only European country to experiment with
inoculation. In the same year, the first inoculations in continental
Europe outside the Ottoman Empire were being carried out by Johann
Adam Reiman in Bohemia.[6] And Princess Caroline's eldest son was
inoculated by Maitland in Hanover in 1724; he suffered over 500 pocks,
but recovered well. Maitland was rewarded with £1,000, and the professor of surgery in Hanover wrote a pamphlet praising inoculation.[7]

A parallel process was taking place on the other side of the Atlantic.
There both the folk-medicine knowledge and the scientific knowledge
were combined in Cotton Mather. In his very different and peculiar
way, Mather's interests were as wide as Sloane's. Born three years after
Sloane, in 1663, he was an American Congregational clergyman,
working as his father's colleague in Boston, with a disconcerting track
record of involvement in the Salem witch trials. His academic creden-
tials were more encouraging. He had entered Harvard at twelve, grad-
uated at fifteen, and was probably the youngest-ever Harvard Fellow.
Yale University can be proud of him too, for it was Cotton Mather,
anxious to promote a Congregational college in Connecticut, who
persuaded Elihu Yale – the Boston-born London merchant who had
become Governor of Madras – to give his money and his name to the
new college. Mather's philanthropy led him in various directions, supporting temperance, education for blacks, Bible Societies, libraries for
working men. When briefly studying medicine in his youth, he had
communicated theories about smallpox to Robert Boyle in London,
and he had become a Fellow of the Royal Society in 1713;* fascinated by the life revealed under the new microscopes, he suggested

---

*He was nominated for the fellowship in 1713, but for some reason the ballot was not taken,
and he was not formally elected until 1723.

that smallpox might be an 'Animalculated Business', caused by tiny organisms in the pustules.[8] His paper to the Society in 1716 had shown him to be enthusiastic for inoculation and, as he had promised, when trouble came he had the courage to pursue his beliefs in practice.

A severe smallpox epidemic struck Boston in 1721 almost at the same time as it struck London. Long aware that port cities were especially vulnerable to the introduction of smallpox, Massachusetts had extended quarantine precautions from ships to the land, passing a smallpox prevention act as early as 1701 which authorised the impressment of houses for the isolation of patients; and Boston had established a quarantine hospital in 1717. But somehow in April 1721 a British ship from the West Indies, the *Sea Horse*, sailed past the quarantine station with two black sailors suffering from smallpox, and others already infected. Great efforts were made, impressive but unsuccessful, to contain the disease: the house where one man lay sick was sealed off, Boston was searched for more cases, twenty-six men were employed cleaning the streets and lanes. The disease spread appallingly, causing three-quarters of all the deaths in the city that year. In June, Mather circulated a manuscript in favour of inoculation round the Boston physicians, describing the accounts from Constantinople in the papers of Timoni and Pylarini and also, as accurately as he could, quoting his slave Onesimus:

> *People take Juice of* Small-Pox; *and cutty skin, and putt in a Drop; then by'nd by a little* sicky, sicky; *then very few little things like* Small-Pox; *and no body die of it; and no body have* Small-Pox *any more.*

Only one of Boston's ten physicians was brave enough to take Mather's advice. Zabdiel Boylston, known for carrying out a successful mastectomy and for his skill in 'cutting for the stone' (an operation dating back to the ancient Egyptians and famously endured by Pepys), inoculated his own six-year-old son, a black slave and a two-year-old black boy. His patients reacted well, but the rest of Boston reacted with fury. 'I never saw the Devil so *let loose* upon any occasion', Mather wrote,

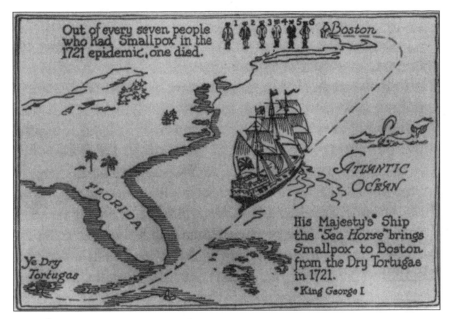

Out of every seven people who had Smallpox in the 1721 epidemic, one died.

Boston

ATLANTIC OCEAN

FLORIDA

Ye Dry Tortugas

His Majesty's* Ship the *Sea Horse* brings Smallpox to Boston from the Dry Tortugas in 1721.

*King George I

12. *The* Sea-Horse *arriving in Boston, 1721*

with an echo of witch-trial rhetoric, 'the people who made the *loudest Cry* ... had a very *Satanic Fury* acting them. They were like the *possess'd People* in the Gospel ... Their common Way was to rail and rave, and wish *Death* or other Mischiefs, to them that practis'd, or favour'd this *devilish Invention*.'[9] Someone threw a grenade into Mather's house, with a note reading, 'Cotton Mather, you Dog. Damn you! I'll inoculate you with this with a pox to you.'

More soberly, a French doctor, Lawrence Dalhonde, gave evidence to the selectmen (Boston's town councillors) about disastrous inoculations he had seen in Italy, and persuaded them that there could be 'most dangerous consequences' in Boston. Arguments flourished in the press, but Boylston went on inoculating – 7 in July, 17 in August, 31 in September, 18 in October, 103 in November. Press opposition became more powerful, with the arrival of the *New England Courant*, the paper run by Benjamin Franklin's elder brother James, which attacked most aspects of the Massachusetts establishment and also aimed 'to oppose the *doubtful* and *dangerous* Practice of *inoculating* the *Small Pox*'. The physician William Douglass (the only physician in Boston with an academic degree in medicine), resenting the intru-sion of non-medical men into medical matters, condemned Mather as a 'credulous vain preacher' and warned (correctly as it turned out) of the danger of infection from inoculation unless the patients were kept isolated until they were fit. 'I reckon it a sin against society to propagate infection by this means', he wrote, 'and bring on my neighbour a dis-temper which might prove fatal, and which, perhaps, he might escape (as many have done) in the ordinary way.'[10] According to a report sent by Henry Newman from Boston to the Royal Society, patients were not kept isolated, but as soon as the eruption began and they felt a bit better 'the Patient *sits* up every Day, and entertains his Friends, yea, ventures upon a *Glass of Wine* with them.'[11]

Newman gave a vivid account of the current methods, describing how, after making incisions:

> *Into these we put bits of* Lint *(the patient at the same time turning his face the other way, and guarding his Nostrils) which have been dipt in some of the* Variolous Matter *taken in a Vial, from the Pustules of one that has the* Small Pox *of the more laudable Sort … Yet we find that the* Variolous Matter *fetched from those, that have the* inocu-lated Small Pox, *altogether as agreeable and effectual as any other. And so we do what is taken from them that have the* Confluent Sort. *[Sloane, who wrote that the matter had to be taken from a 'favourable kind of small-pox' would not have agreed with this alarming advice.] Within Four and Twenty Hours, we throw away the* Lint, *and the* Sores *are dressed once or twice every Four and Twenty Hours, with warmed* Cabbage Leaves.

Carried away with enthusiasm, Newman added:

> *The Patient gets abroad quickly, and is most sensibly* Stronger, *and in* better *Health than he was before … Those that have had ugly* Ulcers *long running upon them, have had them healed … Some very*

> *feeble, crazy,* Consumptive *People, have upon the* Transplanta-
> tion, *grown hearty and got rid of their former Maladies.*

By the end of September, 2,757 people in Boston had been infected with
smallpox and 203 of them had died.[12] Only one inoculated person had
died, and that was probably from an unrelated problem. But when
smallpox and inoculation spread to other New England towns, controversy spread too. Many ministers following Mather's line were attacked
as hypocritical and authoritarian and opposed to the will of God – one
indeed was reported as 'likely to be murdered by an abominable people
that will not let him save his life'. Douglass reminded readers that ministers had persecuted Quakers in 1658, had hanged those suspected of
witchcraft in 1691, and were now just as wrong in 'self-procuring the
Small Pox'. The ministers in their turn defended themselves ('Vindica-
tion of the Ministers of Boston', 1722), and attacked the *Courant* writers as 'Satyrists play[ing] the Divine'. And so it went on, increasingly
ill-tempered. The opposers were 'so unreasonable', wrote Mather, 'that
one had as good speak *Reason* to a *Post*, or argue with a *Whirlwind*'.[13]

The epidemic was over by the following spring, and Mather sent
his account to the Royal Society, defending his actions and hoping to
encourage inoculation in England, from

> *a desire to have our neighbours* do well; *and a solicitude for a* Better
> State of the World. *And all the Obloquies, and Outrage we suffer,
> for our Charity, We shall entertain as* Persecutions *for* A Good
> Cause, *which will not want its Recompenses.*

Starting in June 1721, Boylston himself had carried out 247 inoculations in a year, far more than were being done at that time in England.
Visiting London in 1724 he dedicated his account of inoculation in
Boston to the Princess of Wales, and was rewarded with 1,000 guineas
by King George I.[14]

Others in the New World soon started inoculating both for charitable and for economic reasons – missionaries in the Amazon to protect

the Indians, and plantation owners in the West Indies to preserve their black slaves.

—

In England there were no grenades. The most vocal opposition came from the clergy, led by Edmund Massey, who preached in Christopher Wren's great church of St Andrew's, Holborn, in July 1722 on 'the Dangerous and Sinful Practice of Inoculation'. When Satan 'smote Job with sore boils' he was, Massey asserted, the first inoculator, and inoculation was therefore 'a diabolical operation'. (Maitland mocked in reply that this at least would prove smallpox to be an ancient disease.) But whether or not he managed to convince his hearers about Job, Massey also claimed that smallpox was among 'the wholesome severities ordained for offenders', and that fear of it kept men from vice. It was wrong, too, to interfere with Providence, for:

> The power to inflict disease rests with God alone, and it is He who gives power to heal ... Let the Atheist then, and the Scoffer, the Heathen and Unbeliever, disclaim a dependence upon Providence, dispute the Wisdom of God's Government, and deny Obedience to his laws. Let them Inoculate and be Inoculated, whose Hope is only in and for this Life! But let us, who are better instructed, look higher for Security, and seek principally there for Succour, where we acknowledge Omnipotence.[15]

This was popular stuff, and secular support came from a faction of physicians whose spokesman was William Wagstaffe. A Fellow of the College of Physicians, conservative in his politics and in his medicine, Wagstaffe wrote in his 'Letter to Dr Freind; Shewing the Danger and Uncertainty of Inoculating the Small Pox':

> Some have had the distemper not at all, others to a small degree, others the worst sort, and some have died of it. I have given instances of those who have had it after inoculation in the common way; and consequently

as it is hazardous, so 'twill neither answer the main design of preventing
the distemper for the future. I have considered what the effects may be
of inoculating on an ill habit of body, and how destructive it may prove
to spread a distemper that is contagious: and how widely at length the
authors in this subject disagree among themselves, and how little they
have seen of the practice: — all which seems to me to be just and neces-
sary consequences of these new-fangled notions, as well as convincing
reasons for the disuse of the practice.[16]

He argued that it was wrong to adopt inoculation simply because it
was fashionable; and that it was extraordinary to put faith in this illogi-
cal technique picked up from 'a few *ignorant Women*, amongst an illiter-
ate and unthinking People'. The surgeon Legard Sparham added his
own rhetoric to the debate, saying, 'We have seen South-Sea Schemes,
good Parliaments, Bills for preventing the Plague; heard of plots; but,
till now, never dreamt that Mankind would industriously plot to their
own Ruin, and barter Health for Diseases.'[17]

The pro-inoculation lobby was led by John Arbuthnot who, the
satirist Jonathan Swift said, 'has more wit than we all have, and his
humanity is equal to his wit'. Among his many writings, literary,
mathematical and medical, was 'Mr Maitland's Account of Inoculating the Small-Pox Vindicated from Dr Wagstaffe's Misrepresenta-
tions of That Practice, with Some Remarks on Mr Massey's Sermon'.
Arbuthnot brought statistics as well as rhetoric into the field, using
the London Bills of Mortality to prove that, in arguing that the risks
associated with inoculation were comparable with those of smallpox
caught in the natural way, Wagstaffe had grossly underestimated the
mortality rate of smallpox in London during the previous year. And
so the arguments went on, others joining in on each side, one lot saying
inoculation could not be considered safe until more was known, and
the other saying that it was 'strange to forbid the Practice, till that is
determin'd, which can only be found out by Practice. According to
this Principle, it had been impossible ever to have found out any Thing
in Medicine.'[18]

Bills of Mortality, which were to be appealed to repeatedly in the battles over inoculation, were simply the weekly returns published by the Worshipful Company of Parish Clerks, giving information about baptisms and burials, and sometimes about the causes of death, in each parish. They had been started in the 1530s and continued until 1842, when they were replaced by the Registrar-General's returns. As sources of information about the incidence of diseases they were not ideal, as they were neither inclusive – they excluded burials not under the auspices of the Church of England – nor accurate; in years in which they have been compared with parish registers, the agreement is not good. But they were much better than nothing, and better than most other countries produced. In 1755 Matthew Maty wrote to the British ambassador in Constantinople asking about the number of inhabitants in that town, and how many of them had been carried off by a recent epidemic of plague. He was told that the Turks 'are prohibited by their law to enumerate the people' and that the best way of estimating the number of dead was from the decrease in the consumption of corn – which suggested that 'those either dead or fled amounted to 135,000'. Another way of following the surge in the death rate during the plague epidemic was from the records of the 'colonels of the Janizaries who had their stations at the most noted and only places where the funerals pass'. During twelve days at the peak of the epidemic the number of dead carried out of the Adrianople gate increased just over eightfold compared with the previous year.[19]

The practice of inoculation in England got off to a fairly slow start. Those inoculated came mainly from the upper classes (who could afford the money and the time) or their servants (who were inoculated so that they would not introduce smallpox to their employers' families). It was more convenient if servants had already got over smallpox, as in this 1774 advertisement:

> *Wanted, a man between 20 and 30 years of age, to be a footman and under butler in a great family; he must be of the Church of England and have had the small-pox in the natural way. Also a woman, middle*

aged, to wait upon a young lady of great fashion and fortune; the woman must be of the Church of England, have had the small-pox in the natural way, very sober, steady, and well behaved, and understand dress, getting up lace and fine linen, and doing all things necessary for a young lady that goes into all public places and keeps the best company.[20]

In June 1722 Thomas Nettleton, a physician in Halifax who had tackled a local epidemic of smallpox by inoculating (successfully) more than fifty patients, wrote to James Jurin, the Secretary of the Royal Society, extending Arbuthnot's argument about comparative risks:

I doubt not but when you have collected a sufficient Number of Observations for it, you will be able to demonstrate, That the Hazard in this Method [i.e. inoculation] is very inconsiderable, in proportion to that in the ordinary way by accidental Contagion, so small that it ought not to deter any Body from making use of it.[21]

And he enclosed figures for Halifax, Rochdale and Leeds showing that on average, something like a fifth or a sixth of patients who had caught smallpox during the epidemic had died of it. In a further letter he added:

Whenever any shall happen to miscarry under this Operation [inoculation], that will indeed be very unfortunate & ill, but in this case you will have recourse to the Merchants Logick: state the Account of profitt & Loss to find on which side the balance lyes with respect to the Publick, & form a judgement accordingly.[22]

Jurin, Latin scholar, physician, physiologist and disciple of Newton, responded to the challenge. Using figures collected from inoculators all round the country as well as London, he reckoned that by December 1722 a total of 182 people had been inoculated, with only two fatalities – and even these two may not have been the direct result of the inoculation.[23] By 1724, Jurin had records of effective inoculations in 440 patients in Great Britain, of whom nine had died.[24] Of these nine, two showed signs of smallpox so early that they were thought to have been infected before the inoculation, and the deaths of some of the others may not have been the result of inoculation. Similarly encouraging figures, he said, came from Cotton Mather in Boston where, by March 1722, about 300 had been inoculated and only five had died – again probably not all as a result of the inoculation.[25]

Clearly the risk of being inoculated was much smaller than the risk from catching smallpox 'in the natural way', but not everyone who was inoculated would necessarily have caught smallpox otherwise. So was the risk worth taking? For, as Jurin put it: 'People do not easily come into a Practice in which they apprehend any Hazard, unless they are frightened into it by a greater danger.' He therefore turned to the London Bills of Mortality, choosing the years 1667–86 and 1701–22 because they were years in which deaths from smallpox were recorded as a separate category. And what he found was that, though the percentage of deaths due to smallpox varied from year to year, on average 'upwards of seven per Cent. or somewhat more than a fourteenth part of mankind, die[d] of the Small Pox'. So, from the point of view of the individual being inoculated, he concluded that the risk was certainly worth taking.

The individual point of view was not, though, the only one that had to be considered. On both sides of the Atlantic there was growing concern that widespread inoculation, while protecting individuals, would spread the disease in the community. In 1723, Boston physicians accused Boylston of spreading smallpox through his inoculations – which made the selectmen order him (though ineffectively) to stop. Similar bans, to prevent new outbreaks when the epidemics were over, were tried in Charleston, New York, Connecticut, Maryland and Virginia. Estimating the risk was difficult. The argument against inoculation was that even if, on average, each person inoculated infected only a few people, each of those could pass it on in turn, starting new chains of infection. In England it was suggested that the 50 per cent rise in smallpox deaths in London in 1723 might have been the result of infection following inoculation. But Jurin pointed out that there were 3,271 deaths from smallpox in 1723, so the number of cases must have been about 20,000; since the number of people inoculated in that year was only sixty-eight, the contribution of inoculation to the spread of infec- tion must have been minuscule. The hypothesis that inoculation was responsible for the sharp rise in the incidence of deaths from smallpox in 1723 was, anyway, not very satisfactory, as there had been similarly high incidences on three occasions earlier in the century.

Following his first paper in 1722, Jurin published reports annually till 1727, when he retired as Secretary of the Royal Society. His argu- ments were powerful and influential in England, and were to be quoted a generation later when the same controversy emerged in France.

Why did it take so long for inoculation to reach France? There was no shortage of smallpox — Voltaire suffered from it in a severe epidemic in Paris in 1723 in which, he tells us, 20,000 died. Sir John Vanbrugh gave even more alarming figures in a letter to Lord Carlisle (1724):

> *There has died of the small-pox in Paris this last year 40,000 people, whereas there never died in London in the worst years above 3,000. This has occasioned the physicians at Court there writing over hither to Sir Hans Sloane and Mr Amyand, the King's surgeon (who inoculated the young princesses) to know what success that practice has really had here. The inquiry has been in regard to the King of France. But the priesthood recently stept in, and had the matter sent to the Sorbonne, whose wisdom and piety do not think fit to allow of it.*[26]

It was Dr Jean Delacoste who had written to Sloane for information; but Delacoste, anxious to 'obtain a patent for that practise'[27] and use it to push himself, was not a good advocate, and Wagstaffe's opposition faction had many followers. The deeply conservative doctors of the Paris Faculté as well as the theologians of the Sorbonne resisted all alarming foreign practices. The French Académie Royale de Sciences had not become a centre for scientific thought as the Royal Society had in England – there had been nine articles in favour of inoculation pub-lished in the *Philosophical Transactions of the Royal Society* in 1722 and 1723 alone. The French *Philosophes* were supporters of inoculation, but it was not until 1733, twelve years after the inoculation of Lady Mary's daughter, that Voltaire published his essay 'Sur l'Insertion de la petite vérole'. It was twenty-one years later still that Charles-Marie de la Condamine published his first *Mémoire sur l'Inoculation*.

La Condamine, scientist and traveller, had suffered from smallpox in his youth, and learnt about inoculation when he was in Constanti-nople in 1731–2. Later, while exploring in the Amazon region, sur-veying for the Académie and collecting cinchona seeds as a source of quinine for the fight against malaria, he had heard from Portuguese missionaries how inoculation had saved the lives of groups of Indians. Back in France he wrote his influential *Mémoires* and a two-volume his-tory of smallpox inoculation, telling the story of the practice, tackling the objections and prophesying success. He reckoned that a tenth of the deaths in France were from smallpox, and that it 'destroys, maims, or disfigures the fourth part of mankind'. He also used the utilitarian argument that by saving lives – and he estimated that 25,000 lives a year could be saved in France – the wealth of the state would be increased. In 1760, the Swiss polymath Daniel Bernoulli used the differential cal-culus and the laws of chance to develop a much more sophisticated approach to the analysis of the spread of infections; applied to small-pox his analysis showed that inoculation could bring very substantial benefits, both individually and to the country as a whole.[28]

In France, as in England, the royal family helped to lead the way. There had been a scare when the Dauphin nearly died of smallpox, and not long afterwards, in 1756, the Duke of Orleans, with the rather grudging support of the King, invited Théodore Tronchin to inocu-late his two children. At this stage the Académie weighed in, show-ing approval by awarding Tronchin foreign membership. Tronchin had done much to promote inoculation both in his native Geneva and in Holland, and his success with the duke's children brought great publicity to the cause and great wealth to him. Carriages queued to reach his house, poems and plays were written celebrating inoculation, and the world of fashion responded with '*bonnets à l'inoculation*' (with spotted ribbons) and 'tronchine gowns' (to be worn on health-giving morning walks). Smallpox is also reputed to have influenced fashion earlier through the strange custom of 'patches' to disguise pockmarked faces.

Throughout the controversies, inoculation continued in England and in America. In the 1738 epidemic in Charleston, 800 were inoculated, with only eight deaths. Worried that many might be put off inoculation by the cost, Benjamin Franklin suggested that 'some skilful Physician' should write a do-it-yourself pamphlet to encourage parents to inoculate their own children; in 1759 his friend the English physician William Heberden (who was to become famous for, among other things, his paper distinguishing chickenpox from smallpox) did just that, printing 2,000 copies at his own expense. Franklin distributed them without charge in America, adding an introduction with figures to show the safety and success of inoculation in Boston.[29] There was nothing in the pamphlet about the need for isolation.

In 1743 inoculation was made compulsory for all children at Thomas Coram's new Foundling Hospital, and three years later a severe epidemic in London led to the setting up of the first Smallpox Hospital,

> *for the relief of poor distress'd housekeepers, labourers, servants, and strangers, seiz'd with this unhappy distemper, who will here be immediately reliev'd in the best manner without expence.*

Starting in two houses just south of Pentonville Road, one for inoculation and the other for natural or inoculated smallpox cases, the hospital moved later to a new building at Battle Bridge, the site of the present King's Cross Station (then the edge of London). It

13. The Smallpox Hospital, King's Cross, 1806

was all on a small scale – a total of 3,506 natural smallpox patients and 1,252 inoculated patients in the first ten years. In fact it was on too small a scale to cope with a serious outbreak. In time of crisis, with the best of intentions but with potentially disastrous results, more were inoculated than could be kept isolated in the hospital, and also many of the poor were inoculated in their own homes. So inoculation may well have contributed to the major outbreak of 1752 (when far more were being inoculated than in the epidemic of 1723) – 3,500 died of smallpox that year in London, but since the main risk was from natural smallpox, the outbreak led to more inoculation. St Andrew's, Holborn, became the scene of a very different sermon from the one given by Edmund Massey thirty years earlier. Now Isaac Maddox, in the annual celebrations held for the Smallpox Hospital, gave powerful and well-informed support to inoculation in a sermon which was published, and read all over Europe. It was another great influence on la Condamine and on the movement for inoculation in France.

Smallpox and inoculation were appearing in English literature,

dominating illness and death in the way consumption came to dominate later. Fielding has a pockmarked heroine in *Joseph Andrews*, and
uses smallpox as a convenient way to account for deaths in *Tom Jones*;
Thackeray, in *Henry Esmond* (which is set in the time of Queen Anne),
makes much of the plot turn on the disfigurement by smallpox of his
heroine Lady Castlewood. Later in the century Goldsmith, whose
childhood smallpox had left him badly scarred, gave Mrs Hardcastle,
in *She Stoops to Conquer,* the line: 'I vow, since inoculation began there
is no such thing to be seen as a plain woman; so one must dress a little
particular, or one may escape in the crowd.' Nineteenth-century literary heroines rarely had smallpox, though it is generally assumed to be
the nameless fever that left Esther scarred after she had nursed her sick
maid in Dickens' *Bleak House*; in the Countess of Ségur's story for children *Les petites filles modèles* the roles are reversed and the devoted nurse
catches it from the young heroine.

Turning back from fiction to eighteenth-century letters and diaries,
we often find concern over inoculation. Josiah Wedgwood (founder
of the pottery firm and grandfather of Charles Darwin), who had
himself suffered severely from smallpox as a child, had two children
inoculated and both

> had Convulsions at the first appearance of the eruption [and] have had
> a pretty smart pox as our Doctor terms it. I believe they have had no
> dangerous symptoms, but have been so very ill that I confess I repented
> what we had done, and I much question whether we should have courage
> to repeat the experiment, if we had more subjects for it.[30]

(Later in the year though, the Wedgwoods did have their next baby
inoculated.) Thirty years later the diarist and novelist Fanny Burney
wrote to her father how a successful inoculation had left her 'Relieved
at length from a terror that almost from the birth of my little darling
has hung upon my mind'.[31] Parson Woodforde, in Norfolk, recorded
in 1791 that 'the small-Pox spreads much in the Parish ... It is a pity
that all the Poor in the Parish were not inoculated. I am entirely for it.'[32]

And from the other side of the Atlantic, Benjamin Franklin recalled sadly in his *Autobiography* that:

> *In 1736, I lost one of my sons, a fine boy of four years old, by the smallpox taken in the common way. I long regretted bitterly, and still regret, that I had not given it to him by inoculation. This I mention for the sake of parents who omit that operation on the supposition that they should never forgive themselves if a child died under it — my example showing that the regret may be the same either way, and that therefore the safer should be chosen.*[33]

So in the course of the eighteenth century the practice of inoculation was well established in England, was taking root in France, Holland and Geneva, and had spread patchily in America. In England official approval was given to the cause in 1755 by a statement from the College of Physicians 'That in their Opinion the Objections made at first to it have been refuted by experience, and that it is at present more generally esteemed and Practised in England than ever, and that they Judge it to be a Practice of the utmost benefit to Mankind.'

7

The heyday of inoculation

The physicians approved; but the technique of inoculating, which had been relatively simple when it originated as folk medicine, had become more elaborate and intrusive in their hands. 'In England', it was reported in 1758, 'almost each Physician has a particular Method of preparing his Patients for Inoculation, and many of them make a great Mystery of it.'[1] Mercury, antimony or Peruvian bark (quinine) might be prescribed, with a light diet for two weeks, and a horrible regime of bleeding, vomiting and purging. Lady Mary had complained that 'the miserable gashes that they give people in the arms may endanger the loss of them, and the vast quantity they throw in of that infectious matter may possibly give them the worst kind of small pox';[2] and she suspected that the elaborate preparation was done for the sake of the physicians rather than the patients. This was the time when Edward Jenner, away at school at the age of eight, was made to undergo a particularly long ordeal. The Revd T. D. Fosbroke, who had the account from Jenner himself, says his operation was carried out after

> a thorough preparation which lasted six weeks. He was bled to ascertain whether his blood was fine; was purged repeatedly, till he became emaciated and feeble; was kept on a very low diet, small in quantity; and dosed with a diet drink to sweeten the blood. After this barbarism of human veterinary practice he was removed to one of the then usual inoculation stables and haltered up with others in a terrible state of disease, although no one died.[3]

Poor Jenner might have done better a few years later, when Robert Sutton and his eight sons came on the scene. They reduced the prepara-tory period (which eventually disappeared altogether), and by making far slighter incisions and using serum taken from the developing pus-tules on only the fourth day, while the fluid inside was still clear, they managed to produce a much milder reaction; they also recommended fresh air during convalescence. After a rather shaky start, when 1,200 died of the first 30,000 inoculated by the brothers,[4] Daniel Sutton set up on his own, eventually claiming to have inoculated 40,000 with only five deaths. This was fine for the patients, though, as people were increasingly aware, unless the fresh air was taken well away from the rest of the population there was a danger of spreading the infection. Jenner himself trained in the Suttonian system when he was a young apothecary, and learnt the importance of taking matter from the pus-tules before they matured – a principle he was to use later in his vac-cination technique.

The Suttons were not alone in promoting a gentler approach to inoculation. James Kirkpatrick, who had worked in the Charleston epidemic, advised that 'the smallest violation of the surface of the skin ... was sufficient entrance for the variolous matter to be passed by arm-to-arm transfer'.[5] Théodore Tronchin had his own mild tech-nique – 'to raise a small blister on the arm and to pass through the fluid a thread moistened with smallpox matter'.[6] And Angelo Gatti, a Tuscan doctor who had travelled in North Africa and Greece and who became Tronchin's successor as fashionable inoculator in France, was promising 'the benefits of inoculation without its risks', by spe-cialising in producing very slight reactions – so slight, his opponents said, that they were useless. So when Paris had a severe smallpox epi-demic in 1762–3, he was accused both of failing to protect his patients adequately, and also of letting his inoculated patients spread the infec-tion; further inoculation was forbidden in the city, but allowed in the country outside if the patients were isolated for six weeks. The Paris Faculté, asked to give their view on inoculation, produced two reports, one saying that 'the operation has been and may be attended with fatal effects, and that consequently it ought not to be tolerated', and the other more in tune with the *philosophes* who were welcoming inoculation as the finest medical discovery ever made.[7] The arguments went on for years. Matthew Maty, setting out from Calais in 1764, wrote despair-ingly to the French *Gazette Littéraire*, 'From the shore from which I am going to embark, I cast my glance alternatively on the land where every year more than ten thousand lives are saved by inoculation, and on the one where such a beneficial practice is still rejected.'[8]

This was just when Daniel Sutton, the 'pocky doctor',[9] was start-ing his inoculation centre in Essex, on the road between London and the port of Harwich (the route to the continent and to Hanover, the native principality of the eighteenth-century English kings). Sutton lodged his wealthy patients in houses in the village, where they could stay secluded in the gardens, even with their own parson, until the rashes had cleared. He made huge profits from these patients – 2,000 guineas in his first year, 6,000 in his third – but treated the poor without charge. The problem was that his fame brought crowds – perhaps over a hundred of the poor would come in a day – and these could not all be quarantined. Local inhabitants were alarmed, finding that people were avoiding their markets and their inns for fear of infection, and they even brought an action against him at Chelmsford Assizes, but the Grand Jury threw it out. The attempt was mocked in a pamphlet, 'The Tryal of Mr Daniel Sutton for the High Crime of Preserving the Lives of His Majesty's Liege Subjects by means of Inoculation'. But a letter to the *Gentleman's Magazine* in 1767 put the problem stark-ly: 'The inoculated live: but numbers who receive the infection from them, in the natural way, die.' Nevertheless, Sutton flourished, had a fashionable portrait painted by Joshua Reynolds, and was granted a coat of arms with the motto 'Safely, quickly and pleasantly'. The 'Suttonian system' with its secret medicines (in fact mercury pills) had many converts, and by 1768 there were forty-seven 'authorised' part-nerships (including eight in the Sutton family) in England, Ireland, Wales, Holland, France, Jamaica and Virginia – a highly successful early example of a franchise system.

When Catherine the Great wanted to be inoculated herself and to introduce the practice into Russia, she asked the Russian ambassador in London to recommend the best person. The story is that he suggested Sutton, but that Sutton refused because he was unwilling to take the risk — failure would be so terrible. The next choice was a follower of Sutton's methods, Thomas Dimsdale, who set out in 1768 with his medical student son. The Empress would not let her own physicians share the responsibility by working with Dimsdale but, sympathising with Dimsdale's concern for his own safety if there were a disaster, she had relays of post-horses ready from St Petersburg to the frontier so that he could if necessary flee from her angry subjects. The operation was done in secret, and all went well. As soon as she had recovered Dimsdale inoculated the Grand Duke. As in England, news of the successful royal inoculations encouraged the nobility to follow suit, and the Empress was praised by Senate and Church. Voltaire wrote to her:

> *Oh, Madam, what a lesson Your Majesty is giving to us petty Frenchmen, to our ridiculous Sorbonne and to the argumentative charlatans in our medical schools! You have been inoculated with less fuss than a nun taking an enema ... We French can hardly be inoculated at all, except by decree of the* Parlement. *I do not know what has become of our nation, which used at one time to set great examples in everything ...*[10]

The French were indeed still reluctant inoculators. Arnold Klebs wrote in his history of inoculation (1914) that by 1769, while 200,000 inoculations had been carried out in England, there had been only 15,000 in France.[11] It was another five years before the death of Louis XV from smallpox shocked the French royal family into adopting inoculation and brought it back to Paris. Fashion this time responded with a hair-style, '*pouf à l'inoculation*'.

As well as carrying out inoculations himself Dimsdale, aided by his son, taught the technique to Russian physicians and helped set up inoculation hospitals in St Petersburg and Moscow. He was massively rewarded by Catherine – a fee of £10,000, £2,000 travel expenses, a pension of £500 a year, diamond-studded miniature portraits of herself and the Grand Duke, and the hereditary title of Baron of the Russian Empire.

Dimsdale reckoned, probably with much exaggeration, that in spite of inoculation, 2 million Russians died annually from smallpox.[12] But did inoculation itself, by spreading the infection, contribute to an awful massacre? As so often with smallpox figures, it is impossible to know. Certainly the practice of inoculation soon reached the provinces and big estates, encouraged by the Empress, who believed like la Condamine in the economic importance of a large and healthy population. It is unlikely that there was effective quarantine.

Back home, Dimsdale showed a few years later that he was well aware of the infection problem. He believed that as far as possible all the inhabitants of a village or district should be inoculated at the same time, since 'more lives are now lost in London than before inoculation commenced and the community at large sustains a great loss'. Trouble arose from a philanthropic scheme to set up a Society for the Inoculation of the Poor in their own homes through a free dispensary; when Dimsdale was asked to be consulting physician, with his son as physician, he refused and opposed the whole scheme, because he knew there would be no chance of isolating the patients. The Society went ahead, Dimsdale was furious, and both sides rushed into print. The prosperous Quaker doctor John Lettsom now came into the picture. Concerned that inoculation was generally available only to the rich (and dismissing the need for isolation), he wrote three heated pamphlets attacking Dimsdale and defending free dispensaries. 'It was not', Lettsom's biographer admits, 'a dispute that either of the contestants cared to remember in after years.'[13]

And it was not the first time that Lettsom had offended Dimsdale. Two years earlier there had been trouble over Omiah, the Tahitian brought to London by Captain Cook – Omiah was staying at Dimsdale's house preparing to be inoculated when Lettsom met him at

14. *John Coakley Lettsom, 1792*

dinner and, without consulting his host, wrote up the story of the evening in magazine articles. (The Tate Gallery is currently (2003) hoping to buy Joshua Reynolds' famous portrait of Omiah.)

Lettsom became a prominent figure in the smallpox story, though he is perhaps remembered today only for the anonymous rhyme:

> When any sick to me apply,
> I physics, bleeds and sweats 'em
> If after that they choose to die'
> Why verily! I Lettsom

His parents, small plantation owners in the West Indies, had seven sets of twins, all boys, of whom only the last pair survived. Either to give him a chance of life in a healthier climate, or the chance of a better education, John was sent at the age of six on the two-month journey to England in the care of a Quaker sea captain, to be brought up by the Quaker community in Lancaster, while his brother stayed behind. It was seventeen years before he made his only return visit to his home and family; his father had died, his mother remarried. In England, studious, and helped by the patronage of the influential Quaker family of Fothergills, he flourished. He had a considerable reputation both for his medical skill and for his generosity, and he set up the first General Dispensary in London. So his motives were good, as were Dimsdale's, but in the increasingly bad-tempered exchanges, Dimsdale's arguments were right. A generation later, when Jenner solved the infection problem by using cowpox, Lettsom was one of the first to support him and to acknowledge the dangers of inoculation.

Before this controversy arose, Dimsdale had been asked by Lord Cathcart, the British ambassador to Russia, to help diplomatic relations by calling on Frederick the Great of Prussia on his way back from the mission to the Empress Catherine. But the visit was not a success – the King kept him waiting two hours, then, saying 'Sir, I think you inoculated the Empress and Prince at Petersburg', he wished him a good journey.[14] It was another seven years before Frederick summoned an English inoculator to Berlin to instruct Prussian physicians. So much for one of the patrons of the Bach family. It is more interesting to turn to one of Mozart's.

In 1767 Leopold Mozart took his daughter Nannerl and eleven-year-old Wolfgang to Vienna. Maria Josepha, daughter of the Empress Maria Theresa, was to be married to Ferdinand IV of Naples and Sicily, and with any luck such a celebration should have brought a commission to the young composer. But instead there was disaster – an outbreak of smallpox, with Maria Josepha among the many who died. When the question of inoculation had cropped up in Paris three years earlier, Leopold had refused:

Do you know what people here are always wanting? They are trying to persuade me to let my boy be inoculated with smallpox ... for my part, I leave the matter to the grace of God. It depends on His grace whether He wishes to keep this prodigy of nature in the world in which He has placed it, or to take it to Himself.[15]

Now they were in the thick of an epidemic Leopold felt differently:

In the whole of Vienna, nothing was spoken of except smallpox. If 10 children were on the death register, 9 of them had died from smallpox. You can easily imagine how I felt; whole nights went by without sleep, and I had no peace by day. I decided to go into Moravia just after the death of the princess bride, until the initial sadness in Vienna was in some degree over; only they wouldn't let us go, since His Majesty the emperor spoke about us so often that we were never sure when it might occur to him to summon us: but as soon as the archduchess Elizabeth became ill, I wouldn't let anything hold me up any longer, because I could hardly wait for the moment when I could take my little Wolfgang out of Vienna, which was completely infected with smallpox, into a change of air.[16]

Nannerl and Wolfgang both did catch smallpox, and it is lucky for us all that they recovered.

Rather late in the day, after many deaths in her own family, the Empress Maria Theresa became disillusioned with her Dutch doctor, who believed in the red treatment, bleeding and darkened rooms. She became a convert to the idea of inoculation and, like Catherine of Russia, turned to a disciple of the Suttonian system to save the young archdukes. Jan Ingenhousz, another Dutchman, had been practising in England for the past three years and was recommended by George III's physician. (Later, in 1783, George III himself was to suffer the loss of his four-year-old son Octavius after inoculation.) Besides being a skilled inoculator, Ingenhousz was an impressive scientist, described in one place as a physician and physicist who worked on electricity and magnetism, in another as the biologist and physiologist who discovered photosynthesis and plant respiration. It is not surprising that such a man should have corresponded with Benjamin Franklin on many matters, and indeed he asked Franklin's advice about inoculating the young princes in Vienna.

Before operating on the Hapsburgs, Ingenhousz spent five weeks inoculating poor children, whose parents were paid a ducat for agreeing. By that time, when the infectious matter had been selected from a series of increasingly mild cases, Maria Theresa decided it was safe to go ahead. The success, Leopold Mozart reported, made everyone want to be inoculated.

———

If the eighteenth century became the heyday of inoculation, it was also the heyday of smallpox, with a catalogue of disasters across Europe – in Germany, Poland, Sweden, Denmark, Greenland, Switzerland, Italy, Spain. In 1707, in the most devastating of the nineteen smallpox epidemics that invaded Iceland in the centuries before vaccination, 18,000 died out of a population of only 50,000.[17]

Beyond Europe, too, smallpox seemed to be everywhere. There had already been smallpox in Angola in the seventeenth century. Richard Mead had written of 'natives of that country, who are called Hottentots' in the Cape of Good Hope catching smallpox from doing the washing for sailors in 1713; the washing was done in the Slave Lodge, where 200 of the 570 slaves died.

> But as soon as fatal experience had convinced this ignorant people that the disease was spread by contagion, it appeared that they had natural sagacity enough to defend themselves. For they contrived to draw lines round the infected part of their country, which were so strictly guarded, that if any person attempted to break through them, in order to fly from the infection, he was immediately shot dead.[18]

The slaves also acted as nurses in the white population and nearly every white family was affected – this first epidemic killed a quarter of the Europeans in Cape Town. Ships brought smallpox to the Cape twice more in the century. The first time was a major disaster, killing over 1,000 Europeans and even more native Africans. By the next time, some notification and isolation measures were in place, and ships were inspected so that only the healthy were allowed to land. The disease remained endemic for two years; the Europeans, having in the meantime learnt of variolation, lost only 179, but the native tribesmen suffered more severely.[19]

Cotton Mather's West African slave was familiar with smallpox and with some sort of inoculation, but it is not known when smallpox first arrived in West Africa. There are plenty of stories of epidemics among slaves – at the African ports, at sea (where they were in danger of being thrown overboard to prevent the infection spreading), or bringing smallpox to America. Nineteenth-century explorers described the devastation in African slave caravans and trade routes: Richard Burton saw porters 'staggering on blinded by disease, and mothers carrying on their backs infants as loathsome objects as themselves';[20] David Livingstone, distressed by smallpox that he could not check, tried unsuccessfully to produce vaccine by inoculating a heifer with smallpox;[21] Henry Stanley wrote of an outbreak where fifty to seventy-five were dying daily out a population of 3,000.[22]

Alexander von Humboldt heard that smallpox in parts of South America drove the natives to 'burn their huts, kill their children, and renounce every kind of society'.[23] Similar stories came from the Horn of Africa. James Bruce, travelling in Abyssinia in 1769, wrote a harrowing account of smallpox in Maitsha, near the source of the Blue Nile:

> *When one of these houses is tainted with the disease, their neighbours, who know it will infect the whole colony, surround it in the night, and set fire to it, which is consumed in a minute, whilst the unfortunate people belonging to it (who would endeavour to escape) are unmercifully thrust back, with lances and forks, into the flames, by the hands of their own neighbours and relations, without an instance of one ever being*

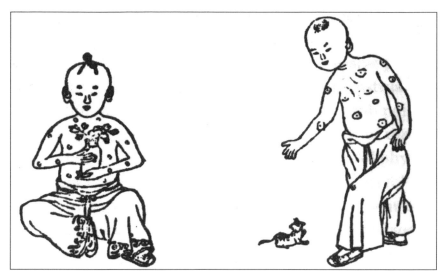

15. Eighteenth-century Chinese illustration of smallpox

suffered to survive. This to us will appear a barbarity scarcely credible:
it would be quite otherwise if we saw the situation of the country under
that dreadful visitation of the small-pox; the plague has nothing in it
so terrible.[24]

And the old centres of smallpox in the East were still full of it. In
1793 Lord Macartney, who was leading the first British embassy to
the Emperor of China (and who was married to Lady Mary Wortley
Montagu's granddaughter), reported home that thousands were dying
of confluent smallpox. The Panchen Lama, persuaded to visit Peking
in 1780 in spite of the risk of smallpox, caught it and died within a few
weeks of arrival; the Dalai Lama built smallpox hospitals in Lhasa.[25]
Holwell wrote of 'malignant confluent' smallpox raging every seventh
year in Bengal. 'The usual resource of the Europeans', he added, 'is to
fly from the settlements and retire into the country before the return of
the smallpox season.'[26] A third of the population of Bengal is said to
have died in the epidemic of 1769–70 – inoculation, more common in
Bengal than in the rest of India, may possibly have helped the infection
to spread.[27]

Fifteen months after the settlers' arrival in Sydney in 1788, an appall-ing epidemic of smallpox attacked the unprotected aborigines of Aus-tralia, as the native Americans had been attacked nearly two centuries earlier. The source of this smallpox has recently been questioned, as the settlers themselves seem to have been free of the disease, and there is a theory that it may have spread from infections brought by Indonesian fishermen to the north coast.[28] Since the middle of the eighteenth cen-tury traders had been coming to collect 'trepang' (sea slugs), valued by the Chinese as food and as an aphrodisiac. Sailing from islands where smallpox was a frequent peril, the fishermen arrived – possibly as many as 3,000 of them annually – with the north-west monsoon, and stayed, processing their catch and mixing with the local population, until the south-east monsoon helped blow them back nearly six months later.

Whatever caused the smallpox outbreak in 1789, there seems to be little doubt that it was the trepang fishermen who took epidemics to Australia in 1830 and again in the 1860s. Smallpox had reached Indo-nesia, probably from China, some time before the Portuguese arrived in the sixteenth century, but it came without the Chinese traditions of inoculation. Java was densely populated, and the disease probably became endemic there, while the other islands suffered the fierce spor-adic epidemics of isolated communities.

Smallpox is a recurring theme in the eighteenth-century history of both North and South America. In the South, there was an endless tale of disasters all over the continent and in the Caribbean islands, with smallpox brought sometimes by traders, sometimes by slaves. George Washington caught it in Barbados in 1751. Inoculation, which la Con-damine had heard about in the Amazon mission settlements, did not spread into the Spanish colonies until late in the century. The smallpox mortality figures are frightening – 17,000 in Córdoba (Argentina) in 1718, 40,000 around Belém (Brazil) in 1750, a quarter of the 6,000 inhabitants of Concepción (Chile) in 1788, 18,000 in Mexico City in 1779.[29] Mexican silver mines and their labour camps became centres of infection, and traders carrying the silver round the country carried smallpox with it.

In New England and Quebec the population – white, black and Indian – were vulnerable to infections brought by increasing trade and travel. Fur-traders went all round the country, soldiers were on the move, and travelling missionaries could carry infection with them. In 1733 a catastrophic outbreak of smallpox in the vulnerable population of Greenland was caused by a traveller returning from Denmark. The years 1755–7 saw the worst smallpox epidemic in the history of Canada. It played a major part in the Seven Years War (1756–63), when smallpox twice hit Nova Scotia, and the British captured Quebec. 'All the scourges are at one and the same time afflicting this poor country', a Frenchman wrote. 'Upon the troubles of war has supervened an epidemic disease which has been intro-duced by the ships that brought the soldiers.'[30] After the French vic-tory at Fort William Henry, Indians who had fought for the French went home laden with trophies, and disastrously took smallpox back with them – they would never help the French again.[31] Shamefully, there were instances of biological warfare on both sides – 'Could it not be contrived to send the smallpox among these disaffected tribes of Indians?' wrote Sir Jeffery Amherst, the British commander-in-chief in North America. 'I will try to inoculate the — with some blankets that may fall into their hands,' his colonel replied; whether he did so is not known.[32] It was not the only case of the kind – the captain in charge at Fort Pitt recorded in his diary that he had sent the Delaware Indians 'two Blankets and a Handkerchief out of the Small Pox Hospital. I hope', he added, 'it will have the desired effect.'[33] And the French commander, Montcalm, is said to have shipped infected British prisoners of war to Halifax – but the plot misfired, for the guards got smallpox and died, while the prisoners recovered and took over the ship.[34]

In the period of peace before the War of Independence, in some parts of the country inoculation was promoted by the press. The *Gazette du Canada* tried to encourage an unenthusiastic population with a letter

from the Chaplain of Quebec quoting la Condamine.[35] In Pennsylvania, Philadelphia's doctors and wealthy families became enthusiastic followers of Benjamin Franklin's propaganda in the *Pennsylvania
Gazette*. It was partly under Benjamin Franklin's influence, too, that
arrangements were first made for inoculating the poor, in Boston in the
epidemic of 1764 and in Philadelphia ten years later.

Howard Simpson has described one strange and unfortunate side
effect of the smallpox outbreak in Boston in the winter of 1763/4. The
General Court, he says, fled over the river to Cambridge and 'required
such roaring fires in the fireplaces of Harvard Hall that the whole
building burned down, destroying the best collection of scientific
apparatus and the finest library in the American colonies'.[36]

Soon smallpox was to play a part in the mainstream of North
American history, dominating strategy and affecting the course of battles in the War of Independence. 'I know that it is more destructive
to an army in the natural way than the sword', Washington wrote to
Patrick Henry, Governor of Virginia, in 1777.[37] After the Battle of
Bunker Hill in 1775, Washington's worries about smallpox among the
British troops and among the citizens of Boston led him to keep his
army outside the town for nine months of stalemate, until the British
left. He heard a rumour from a sailor that civilians leaving the town had
'been inoculated with the design of spreading the smallpox throughout
the country and the camp'.[38] If true, this would be a strange perversion
of inoculation; but Washington, like many Americans, had an acute
awareness of the danger of inoculated smallpox. He cautiously ordered
that 'as the enemy has with malicious assiduity spread the smallpox
through all parts of the Town, no officer or soldier may go into Boston
when the enemy evacuates the Town'; he then allowed 'a thousand
men who had had the smallpox' to move in. In spite of these efforts,
crowds poured back, the epidemic spread, and the legislature decided
to inoculate vulnerable troops and, for a brief period, to allow general
inoculation. Inoculation became 'as modish as running away from the
Troops of a barbarous George was the last year'.[39] Social gatherings
for inoculation in Massachusetts remind us of Lady Mary's Constan

tinople – John Hancock (the statesman who would later be the first to sign the Declaration of Independence) wrote to George Washington that 'Mrs Hancock would esteem it to have Mrs Washington take the smallpox in her house'.[40]

In 1776 the governor of Virginia assembled a hopeful 'Ethiopian army' of nearly 1,000 slaves, promising them their freedom if they fought for the King. Crowded together in camps or on board ship and vulnerable to disease, 'dozens died daily from Small Pox and rotten Fevers' and their enthusiasm ended in tragedy. It was a tragedy that was to be repeated later in the war. 'By the time the British surrendered at Yorktown in 1781', Elizabeth Fenn has written, 'the virus had dashed the dreams of thousands of freedom-loving African Americans.'[41] In the chaos of war many pox-covered black loyalists were abandoned.

Smallpox took charge of affairs in Quebec in the summer of 1776, when a besieging colonial force 2,000 strong collapsed with 900 sick. 'Our misfortunes in Canada are enough to melt the heart of stone', John Adams, the future American President, wrote. 'The smallpox is ten times more terrible than the British, Canadians and Indians together. This was the cause of our precipitate retreat from Quebec.' And, it has been claimed, the main cause of the preservation of Canada to the British Empire.[42]

Some American states, including George Washington's Virginia and most of New England, banned inoculation for fear of spreading the infection. And some soldiers found their own way round the ban – as explained by an old soldier when he was applying for a pension:

> *While at Quebec, this declarant had the smallpox. He inoculated himself; got the infection from the hospital. He also inoculated many of his soldiers, but as this was against orders, they were sent into his room blindfolded, were inoculated, and sent out in the same condition. Many lives were saved by this measure, as none thus inoculated died, while three out of four who took it in the natural way died.*[43]

There were also more rumours of deliberate infection – of infected prisoners sent back, or infected women sent out to the American camps. It is encouraging to find, though, that the British commander at Quebec organised the rescue of between 200 and 500 abandoned and pox-ridden Americans, sending them off for hospital care.[44]

Eventually the outbreaks of smallpox, and the shortage of recruits (scared off by those outbreaks), persuaded Washington that the policy had to change. 'Finding the smallpox to be spreading much, and fearing that no precaution can prevent it from running through the whole of our Army,' he wrote, 'I have determined that the troops shall be inoculated ... should the disorder infect the Army in this natural way and rage with its usual virulence we should have more to dread from it than from the sword of the enemy.'[45] It was a major undertaking, and it was done as secretly as possible so that the British did not exploit the temporary illness of so many soldiers. But it was done successfully, and Washington no longer had to fear that smallpox would cripple his army.

Shortly afterwards, though, it crippled his allies' navy. In 1779 the French and Spaniards, joining in the war against Britain, planned an invasion across the English Channel with a large combined fleet – 'sailing in double column they covered four and a half miles of sea ... the most powerful armada that ever walked the waters',[46] twice as many ships as the British could muster to oppose them. But the French had some cases of smallpox on board, and the hot crowded ships became floating hospitals. More than half the crews were ill, and so many dead were thrown overboard that, it is said, the people of Plymouth ate no fish for a month. With the wind turning against them as well, the ships scattered and the invasion threat was over.

In 1781 smallpox was attacking Indian tribes as far apart as California and the trading posts of Hudson's Bay. In that final year of the War of Independence Andrew Jackson (then aged fourteen) was taken prisoner, together with his brother Robert, and forced to march forty-five miles to a terrible military prison where they both caught smallpox. Robert died, while Andrew, who became President of the United States in 1828, was permanently scarred with hatred of the British.[47]

With thousands still dying of smallpox all over the world, how suc/ cessful was inoculation, and what hope did it give for the future? From the individual's point of view it was a risk well worth taking, for the chance of a fatal or even a bad reaction was slight; there would prob/ ably be only a mild illness, followed by permanent protection from a dangerous one. It is more difficult to assess the value of inoculation to the community. It could certainly be effective in small populations if used thoroughly and in conjunction with isolation as soon as there was a sign of the disease – the Suttons had examples of total success with mass inoculations in the English towns of Diss and Maidstone. And James Boswell, travelling in the Hebrides in 1773, noted that 'the Laird said he had seven score of souls upon it [the island of Muck]. Last year he had eighty persons inoculated, mostly children, but some of them eighteen years of age. He agreed with the surgeon to come and do it, at half a crown a head.'[48] But what about large towns, and in particular, since it is a little easier to get some sort of information here, what about London? Even in London there are many factors to take into account, and few figures.

Using the Bills of Mortality, Charles Creighton in his *History of Epidemics* (1894), claiming that 'the ordinary course of smallpox in Britain was little touched by inoculation'[49] showed how the annual total of smallpox deaths in London, see/sawing as epidemics came and went, was at least as high at the end of the eighteenth century as at the beginning. But that was perhaps not surprising in an increasingly mobile and expanding population – London grew from 575,000 in 1700 to 900,000 by 1800[50] – and there is no way of knowing whether the figures for smallpox deaths would have been higher still if there had been no inoculation. There was a slight rise in the proportion of deaths in London that were caused by smallpox, but it has been sug/ gested that this was because of rural immigrants who had not grown up, as those born in London had, in a community where smallpox was endemic. Certainly the higher smallpox death toll was among the poor, for although smallpox attacked 'prince and peasant alike',[51] the peasants were less likely to be protected by inoculation. And the Small-pox Hospital with its economy-class inoculation was far too small to keep all patients isolated; 'whoever applied at their gates were inoculated, and suffered to wander through the city of London covered with pustules and exhaling infectious vapours'.[52]

But inoculation was certainly successful enough to inspire visions of eventual victory over smallpox. As early as 1767, Matthew Maty, in his paper 'The Advantages of Early Inoculation', imagined a time when inoculation had been made universal, and so:

When once all the adults susceptible of the infection should either have received it or be dead without suffering from it, the very want of the variolous matter would put a stop to both the natural and artificial smallpox. Inoculation would then cease to be necessary, and therefore be laid aside.[53]

The idea was carried further by John Haygarth, a physician who had long been concerned with the problem of infectious fevers – he was the first to treat fever patients in separate hospitals. In his 'Sketch of a plan to exterminate the casual smallpox from Great Britain and to introduce general inoculation',[54] he pointed out that inoculation was 'still so far confined to the superior and least numerous classes of society ... [and smallpox] continues to be the most fatal malady that ever afflicted mankind'. Because inoculation had been 'eminently useful to the rich' there had been, he argued, a tendency to ignore old precautions, so it had been 'injurious to the poor'. Inoculation had to be continued, especially in large towns, for the poor as well as the rich; it was the *partial* adoption of inoculation in London that caused deaths to rise.[55] So his plan involved inoculation combined with compulsory isolation. This was easier in a well-defined community, and he presented his own efforts in Chester as an example. Clergymen, he wrote, have sometimes been helpful with propaganda for inoculation. Parson Woodforde had declared himself 'entirely for it' two years earlier,[56] and Haygarth quoted Lady Mary's grandson, The Hon. Revd William Stuart, Rector of Luton:

> *Towards the end of last summer, a small pox of the most malignant kind prevailed at Luton … I endeavoured to overcome the prejudice and fears of the people, and to prevail on them to be inoculated. This infection was accordingly communicated to twelve hundred and fifteen, of whom only five died; and soon after, to seven hundred more, with equal success …*

This effort to get all vulnerable members of a community inoculated was exactly what Haygarth was urging. He calculated that

> *if seven out of nineteen hundred and fifteen died of inoculation, and if we reckon the deaths by the casual smallpox at one in five (though the malignity of the epidemic warrants a much higher estimate), it follows that three hundred and seventy six lives were saved by the rector of the parish.*

(The calculation assumes that all those inoculated would otherwise have caught the disease in the natural way.)

Haygarth's advice was to pray regularly for the extermination of smallpox, and to try isolation, cleanliness and care with treating infected clothes. To tackle the problem of compulsory inoculation and isolation 'in such a *free* and *commercial* nation', he proposed a hierarchy of inspectors, commissioners, and directors, with a system of fines, perhaps a shilling or more a day, for transgression, which would help to pay for rewards for the poor who complied. (This, when skilled workers in the textile trade were lucky to get seven shillings and sixpence a week, was not trivial.[57]) Any who would not pay should 'be exposed in the nearest market town, for an hour, with this label on his breast, "Behold a villain, who has wilfully,

16. John Haygarth

and wickedly spread the poison of the small-pox.'" 'If *all* concerned, both officers and people, would perform their duty *exactly,*'
he added, 'the smallpox might be exterminated out of the island in
a few weeks!' He pleaded that this should not be dismissed as 'a
visionary scheme' or 'an extravagant and dangerous innovation'.

Perhaps that is just what it was. But five years later Edward Jenner
published his 'Inquiry into the Causes and Effects of the Variolae Vaccinae', with news of a revolutionary technique that involved no risk of
spreading smallpox. The whole situation changed.

8

From cuckoos to cowpox

… among those whom in the country I was frequently called upon to
inoculate, many resisted every effort to give them the smallpox. These
patients I found had undergone a disease they called the cow-pox,
contracted by milking cows affected with a peculiar eruption on their teats.

Edward Jenner, The Origin of the Vaccine Inoculation, 1801

In 1798 Malthus published his *Essay on the Principle of Population*, and
Wordsworth and Coleridge published *Lyrical Ballads*. In the same
year there was 'printed for the author' (and at the author's expense) a
revolutionary pamphlet by Jenner called 'An Inquiry into the Causes
and Effects of the Variolae Vaccinae, a Disease discovered in some of
the Western Counties of England, particularly Gloucestershire, and
known by the name of the Cow Pox'. According to John Baron, Jenner's friend and biographer, the President of the Royal Society (Sir
Joseph Banks) had advised Jenner not to submit it to them as he 'ought
not to risk his reputation by presenting to the learned body anything
which appeared so much at variance with established knowledge'.[1]

For Edward Jenner was already a Fellow of the Society, with a
reputation as a naturalist and a paper to his credit on the behaviour of
cuckoos. And long before Darwin raised the profile of earthworms,
Jenner described how they broke up the soil and showed, as Humphry
Davy said, how they were 'an agent important to man in the economy
of nature'.[2] It was a great time for amateurs, when enthusiasts, often
doctors or clergymen could, in the tradition of Sloane, collect and

17. Edward Jenner

observe and make significant contributions to the knowledge of the world around them – the Revd Gilbert White with his *Natural History and Antiquities of Selborne*, for example; Dr Lettsom with his botanical collection, his study of bees and his dissertation on tea; or Jenner's friend Dr Caleb Parry (the physician Jane Austen knew in Bath), with his sheep-breeding experiments and his collection of minerals and fossils. Jenner's childhood passion for natural history, encouraged and directed later by his teacher John Hunter, made him a part of this scene.

Born in 1749, Edward Jenner was the sixth, and last, surviving child of the prosperous vicar of Berkeley, Gloucestershire. He seemed to have an idyllic background in the village that was to be his home all his life, but his happiness was shattered when his mother died after giving birth to a son who lived for only a day, and his father died two months later. So at the age of five Edward was left in the care of siblings who, trying for the best, sent him away at eight to the school where he suffered so severely from the inoculation. He left shortly afterwards, and went for two or three years to Cirencester, joining a group of students at the grammar school under the Revd Dr Washbourne. Dr Washbourne failed in his efforts to teach the boy Latin and Greek, but Jenner spent his time observantly roaming the countryside, and starting his life-long interest in fossils; it was probably also while there that he first came across the young Caleb Parry (born 1755), son of a local dissenting minister. Jenner left school clearly not suited to following his brothers in their paths to Oxford and the Church. Someone had the imagina-tion to suggest that a medical career might do, and at thirteen he was apprenticed for six years to a local surgeon, John Ludlow. At the end of that time, according to his friend James Moore, 'he learnt that there was a report, rife in the dairies, of a distemper named the Cow Pox, which infected the hands of the milkers, being sometimes a preventive of the Small Pox'.[3]

For many years Jenner kept this knowledge stored away, while he continued his studies in London. He enrolled in 1770, at £100 a year including board and lodging, as one of John Hunter's first pupils, and he also followed William Hunter's 'full course of Anatomy,

Physiology, and Midwifery'[4] in the new Anatomy School in Windmill Street. From either a personal or a professional point of view, he could not have chosen better. The Hunter brothers were passionate collectors (their anatomical and natural history collections are in the Hunterian Museums in the Royal College of Surgeons and in Glasgow) and great surgeons and teachers. John Hunter had himself as a boy been more involved in the countryside around him than in school-work, turning to medicine under the influence of his older brother. But the fascina-tion with curiosities of nature stayed with him all his life. He bought two acres of land outside London, at Earls Court, where he kept leo-pards, buffaloes, ostriches, jackals and snakes, as well as studying bees, fish and oysters; he bought dead animals from the menagerie at the Tower of London so that he could dissect them. The best rooms in his house, according to his brother-in-law Everard Home, were filled with his museum. His wife Anne added a fashionable literary and musical side to the household – Haydn, who wrote settings for some of her verses, later became part of her circle – a feature which would have been appreciated by Jenner, who enjoyed singing and was a competent player on the flute and the violin.

Jenner remained with Hunter for two years. His fellow pupils included two future presidents of the College of Surgeons – Henry Cline, who became the first man in London to use cowpox vaccine, and Everard Home; both became Jenner's life-long friends. During this time Joseph Banks came back from Captain Cook's first three-year voyage of scientific exploration and, on Hunter's recommendation, Jenner was given the job of cataloguing his botanic collection. Banks was impressed and invited Jenner to join him on Cook's next voyage, to be one of an entourage that grew so large – fifteen people, including the painter Zoffany and a horn player – that the boat could not accom-modate them all. Banks then withdrew altogether, but in any case Jenner had no such ambitions. He never travelled away from southern England, he hated London, and later turned down both the chance of a well-paid appointment in India and an offer of partnership with Hunter. All he wanted was to be a country doctor in Berkeley, and from 1794 also in the summer season in Cheltenham. He became the most famous country doctor in the world.

Back home, Jenner easily built up a local practice and enjoyed a social life of visits, music, and meetings of a couple of dining clubs of medical men, including Caleb Parry and Daniel Ludlow, the son of his first teacher. Jenner contributed various papers, including one on ophthalmia and one, which was written up by Parry, on angina. We have a description of him at that time as he first appeared to his friend Edward Gardner: 'He was dressed in a blue coat, and yellow buttons, buckskins, well-polished jockey boots with handsome silver spurs, and he carried a smart whip with a silver handle. His hair, after the fashion of the times, was done up in a club, and he wore a broad-brimmed hat.'[5] In one notable venture, only a few months after the Montgolfier brothers made their first balloon ascent, Jenner organised an unmanned flight from the grounds of nearby Berkeley Castle – probably only the second such flight in England.

Jenner was fascinated by hibernation. He studied hedgehogs, and thought critically about current ideas that birds might hibernate like bats, in cracks or holes in rocks and buildings, or even in mud at the bottom of ponds; he tested the weight and the stomach contents of birds in springtime, showing that, unlike hedgehogs, they were fat, and were indeed well fed with foreign seeds, which must have been the result of migration. There are many surviving letters from Hunter at that time, with advice about patients mixed up with queries about the habits of various forms of wildlife, or requests for field specimens that might be added to Hunter's museum. 'If you can pick me up anything that is curious and prepare it for me, do it, either in the flesh or fish way', Hunter wrote, suggesting at other times that Jenner could try catching bats and taking their temperature, examine eels to determine their sex and method of propagation, or 'remove the cuckows egg into another bird's nest and tame the young one to see what note it has; there is employment for you, young man ... I want a nest with an Egg in it also a nest with a young cuckow, and also a young cuckow. I hear you saying there is no end to your wants.'[6]

Jenner started to write papers, 'Cursory Observations on Emetic Tartar' in 1783 (a method to improve a standard medicine, which Hunter suggested he should patent), and four years later his 'Observations on the Cuckoo'. The cuckoo's habit of laying eggs in other birds' nests was well known, but what Jenner noticed was that it was the new-born cuckoo chick, not the foster parents, that threw the other babies out of the nest. This seemed so unlikely that even the article on Jenner in the *Dictionary of National Biography*[7] dismissed the idea as absurd, probably the result of fudged reports from Jenner's lazy nephew Henry, who was helping the observations. But Henry had been perfectly conscientious, going the rounds of hedge sparrows' nests containing cuckoo eggs on a scale hard to imagine now that cuckoos are so scarce in England. The truth of the theory was proved by photography early in the twentieth century.

As a country doctor, Jenner practised surgery, general medicine, midwifery and dispensing. He also inoculated, using the improved Suttonian system which, with its milder reactions, encouraged more patients. In this farming community Jenner came across some who had no reaction at all and who told him that they had at some stage in their lives caught cowpox by milking diseased cows. He remembered hearing such stories when he had been an apprentice, and he must have known that his neighbour Mr Fewster had written a paper (never published) back in 1765, on 'Cow Pox and its ability to Prevent Smallpox'. It was a topic which, John Baron claims, Jenner and Fewster talked about so much at his 'Convivio-Medical Society' that 'they threatened to expel him if he continued to harass them with so unprofitable a subject'.[8] Fewster, though, showed no enthusiasm for following up the idea, writing that 'inoculation for the smallpox seems so well understood, that there is very little need for a substitute'. 'It is curious, however,' he added, 'and may lead to improvements'.[9]

As early as 1754 (when Jenner was five years old), Théodore Tronchin had heard stories of the Gloucestershire milkers and their immunity to smallpox, but had dismissed the idea with the remark, 'How can they be so superstitious?'[10] Joseph Adams, who became physician to the Smallpox Hospital, had discussed cowpox in his 1795 book on *Morbid Poisons*, and had written cautiously that 'as far as facts have hitherto been ascertained, the person who has been infected is rendered insensible to the variolous poison'. Back in the fifteenth century the Chinese had had a dubious idea about using pills made from dried and pounded water-buffalo lice (possibly from buffaloes infected with something like cowpox) to prevent smallpox.[11] There are elusive stories from other countries of Europe, and even from Mexico and Persia, about milkers who had 'sore hands' from cowpox and were known to be safe from smallpox.[12] And the Hindu god Krishna is said to have loved milkmaids because of their beautiful unscarred complexions. In 1801 Lettsom summed up the situation prosaically: 'This preventive quality of the vaccine fluid was certainly known even to scientific professional men many years ago; but, strange as it may now appear, no one, till Jenner promulgated his discovery, had ever improved that knowledge by applying it to the process of inoculation.'[13]

So these ideas were around, and Jenner has explained how he thought about them and worried out his theories until he was ready to experiment. With all modern scientific knowledge and techniques, we still do not know the original source of the virus that causes smallpox, though it is likely to have been derived from a virus in animals. Jenner, in the late eighteenth century with only folk tales and rumour to guide him, was already speculating on the way diseases in man might have arisen from the domestication of animals:

> There is a disease to which the Horse, from his state of domestication, is frequently subject. The Farriers have termed it the Grease. It is an inflammation and swelling in the heel, from which issues 'matter' possessing properties of a very peculiar kind, which seems capable of generating a disease in the Human Body ... which bears so strong a resemblance to the Small Pox, that I think it highly probable it may be the source of that disease.[14]

Men dressing the infected heels and then milking cows without 'paying due attention to cleanliness' could, he thought, infect the cows' nipples, giving them cowpox, which in turn could pass to other milkers. And 'what renders the Cow-Pox virus so extremely singular', he added, 'is, that the person who has been thus affected is for ever after secure from the infection of the Small-Pox'.

But, as Hunter had written to Jenner in 1775 about a question of the hibernation of hedgehogs, 'why think — why not trie the Expt.?'[15] The advice echoed Cotton Mather: 'Of what Significancy are most of our *Speculations*? EXPERIENCE! EXPERIENCE! 'tis to THEE that the Matter must be referr'd after all.'[16] Jenner collected his evidence, chronicling (in his 'Inquiry') twelve cases of subjects who had suffered from cowpox and three who had suffered from 'grease'; the last of those three developed a mild attack of smallpox twenty years later, but all the rest had proved resistant to smallpox, whether it was natural or inoculated. If cowpox caught from the teat of a cow protected farm workers from smallpox, seemingly without serious side-effects and without providing a focus for the further spread of the cowpox, could the same safe protection be induced by using cowpox for inoculation? He was ready to do the crucial experiments.

On 19 July 1796 Jenner wrote to his friend Edward Gardner:

> I have at length accomplish'd what I have been so long waiting for, the passing of the vaccine Virus from one human being to another by the ordinary mode of Inoculation.
>
> A boy of the name of Phipps was inoculated in the arm from a Pustule on the hand of a young Woman who was infected by her Masters Cows. Having never seen the disease but in its casual way before, that is, when communicated from the Cow to the hand of the Milker, I was astonish'd at the close resemblance of the Pustules in some of their stages to the variolous Pustules. But now listen to the most delightful part of my Story. The Boy has since been inoculated for the small pox which as I ventured to predict produc'd no effect.[17]

The Jenner museum at Berkeley displays the horns of the cow whose

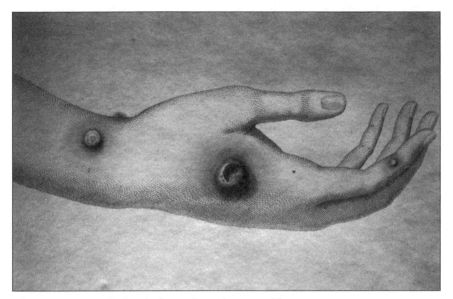

18. Cowpox on the hand of Sarah Nelmes, used by Jenner to vaccinate James Phipps

cowpox infected the hand of Sarah Nelmes when she was milking. In Jenner's 'Inquiry' there is a drawing of the pustule, on her hand, which was used to inoculate eight-year-old James Phipps. The important points were that Jenner deliberately inoculated the boy with cowpox (transferred person-to-person), demonstrated that the reaction made him immune to inoculated smallpox, and understood the significance of what he was doing. Also, by following the Suttons in doing away with the awful preparatory weakening medication, and by producing a far milder and usually uninfectious reaction, he made the process quicker, cheaper and less stressful.

In *The Times* of 4 December 2002, Mark Henderson, while calling Jenner's great experiment 'one of the most significant breakthroughs in the treatment and prevention of infectious disease', added, with anachronistic political correctness, that it was 'achieved by medical malpractice'. Before the days of 'informed consent' and 'ethical committees' this is grossly unfair. Cowpox was a known, transient and relatively mild disease, and by exposure to it Phipps, as Jenner had good reason to believe, both achieved immunity from smallpox and avoided the unpleasant reaction and greater risk involved in inoculation with smallpox (variolation). If Jenner had been wrong, Phipps would merely have had an unnecessary attack of cowpox. More questionable was Jenner's later experiment, designed to prove that cowpox could be caught only through a break in the skin. He 'suffered children two or three times a day to inhale by the mouth and nostrils the effluvia of pustules on the arms of others, when the matter has been in its most active state, and the pustules punctured in several places to give the matter its fullest effect'.[18] But convinced that cowpox, unlike smallpox, could not be caught by inhalation, he saw no risk; and the children were not infected.

When the success of vaccination (the term comes from *vacca,* the Latin for cow, and was introduced in 1803 by Jenner's friend Richard Dunning) had made Jenner famous, others were produced who claimed to have inoculated with cowpox before the experiment on Phipps. Most of these faded away on investigation, but the most substantial was Benjamin Jesty, a farmer in Dorset who, in 1774, had tried to protect his wife and two children from a local smallpox epidemic by scratching their arms and rubbing in matter from cowpox pustules in a neighbouring field. He was not bothered about himself, having had cowpox earlier. The children got over it easily, but his wife became so ill that he had to call a doctor and confess what he had done; she recovered, but the villagers were appalled and abusive, and Jesty only wanted to forget the whole affair. Until, that is, he was produced by detractors of Jenner. His tombstone, forgivingly put up by his wife, states cautiously that he was 'the first person (known) who introduced the cow-pox inoculation, and who, for his great strength of mind, made the experiment from the cow on his wife and two sons'. Edgar Crookshank, trying to play down Jenner's importance, used a portrait of Jesty as the frontispiece to his massive *History and Pathology of Vaccination* in 1889. But Jesty, even though he had tried this one experiment twenty years before Jenner, had not started to collect further evidence about cowpox or to make further trials; and he did not challenge the immunity by a smallpox inoculation – it was fifteen years before his family were routinely inoculated and failed to react.

Jenner, on the other hand, was so anxious to tell the world what he had done and what hope it gave for the future, that he wrote it up as soon as he could – and was criticised for not doing enough experiments to prove something 'which appeared so much at variance with estab-lished knowledge'.[19] Everard Home, in his 'referee's report' to Joseph Banks on the first draft of the 'Inquiry', commented that 'if 20 or 30 children were inoculated for the Cow pox and afterwards for the Small pox without taking it, I might be led to change my opinion, at present however I want faith'.[20] That was the letter behind Joseph Banks's advice to Jenner not to submit the 'Inquiry' to the Royal Society. So Jenner waited two years for more cowpox and to collect more evidence, and then published his findings himself in June 1798, with a dedica-tion to Caleb Parry; eighteen months later, the success of his discovery gave him the confidence to dedicate the second edition to the King.

In the pamphlet, Jenner introduces the term *variolae vaccinae* (small-pox of the cow) and discusses twenty-three 'cases'. The six cases after Phipps (two of them concerning several patients) all involved deliber-ate vaccinations, the first with 'grease' (an unfortunate boy who died from 'a contagious fever in a workhouse soon after') and the rest with cowpox, transferred arm-to-arm from one child to the next. 'These experiments afforded me much satisfaction', Jenner wrote. 'They proved that the matter in passing from one human subject to another, through five gradations, lost none of its original properties.'[21] Indeed since the virus was reproducing itself in each successive subject, and was not being used up, it could with care be passed indefinitely. Jenner challenged three of his cases by variolation. He claimed that vaccina-tion did not cause an eruptive reaction (except at the point at which matter was inserted into the skin) and, above all, was not itself infec-tious. It simply caused 'the usual slight symptoms' – mild fever and local inflammation. He hoped, he wrote, that people would think about his conclusions – today it might almost be called a 'consultation document':

Thus far have I proceeded in an inquiry, founded, as it must appear, on the basis of experience in which, however, conjecture has been occasionally admitted in order to present to persons well situated for such discussions, objects for a more minute investigation. In the mean time I shall myself continue to prosecute this inquiry, encouraged by the hope of its becoming essentially beneficial to mankind.[22]

Jenner took the manuscript of his pamphlet to London to arrange for its publication, and also took with him a supply of vaccine from the arm of one of his patients, anxious to get a chance to demonstrate its virtues. But John Hunter had been dead five years, and no one showed any interest. After three months Jenner returned disappointed to Berkeley, leaving the dried vaccine with his old fellow-student Henry Cline. He had barely been back two weeks though when a letter came from Cline, who had experimentally vaccinated a boy suffering from an inflamed hip, with the idea that the counter-irritation would help. The hip was not helped, but Cline decided to test the cowpox by variolating the boy:[23]

The cow-pox experiment has succeeded admirably ... Dr Lister, who was formerly physician to the smallpox hospital, attended the child with me and he is convinced that it is not possible to give him the smallpox. I think the substituting of the cow-pox poison for the smallpox promises to be one of the greatest improvements that has ever been made in medicine.[24]

Henry Cline's enthusiasm gave Jenner plenty of publicity, and matters developed fast. Jenner's 'Inquiry' had been published on 17 September 1798; in November George Pearson, physician at St George's Hospital, followed it up with his own 'Inquiry concerning the History of the Cow Pox. Principally with a view to Supersede and Extinguish the Small Pox'. Pearson had been busy. Giving full acknowledgement to Jenner and disclaiming any originality, he was supporting the practice of vaccination by publishing the result of a questionnaire sent round to medical men to collect information on the immunity to smallpox of those who had caught natural cowpox. It was a summary of current knowledge, with no pretence of anything new. He variolated five people, three of whom had had cowpox and failed to react, but had not himself yet seen any vaccinations. He concluded that 'The fortunes of the new proposed practice cannot with certainty be told at the present by the most discerning mind, more instances are required to establish practical and pathological truth.'

Pearson was very ready to collect more instances, and was pleased when sources of cowpox were found in a dairy in Gray's Inn Lane and in a herd of cows in Marylebone Fields. Together with William Woodville, the medical director of the Smallpox Hospital, he set to work, in full communication with Jenner, and sending him a bit of thread infected with the vaccine for his own use. But the relationship soured when Pearson started to appear more as a usurper than a supporter. In March 1799 Jenner's nephew George warned his uncle:

> *Dr Pearson is going to send circular letters to the medical gentlemen to let them know that he will supply them with cow-pox matter upon their application to him, by which means he will be the chief person known in the business, and consequently deprive you of that merit, or at least a great share of it, which is so justly your due.*

Pearson was in London, at the centre of medical interest. Jenner clearly needed to be on the scene too, and he came to London again for three months while he published his 'Further Observations on the Variolae Vaccinae or Cow Pox' – a rather hurried explanation of the importance of vaccinating correctly. Opponents of vaccination had been quick to point out its failures. Jan Ingenhousz (who had inoculated the Hapsburgs) had written to tell Jenner of a farmer who had had cowpox and subsequently reacted badly to a smallpox inoculation and fatally infected his father.[25] 'Brickbats and hostile weapons of every sort are flying thick around me', Jenner wrote to Edward Gardner:

> but with very little aid, and a few friendly opiates, seasonably adminis-
> tered, they will do me no injury ... We must set off by impressing the
> idea that there will be no end to cavil and controversy until it be defined
> with precision what is, and what is not Cow Pox. The true has many
> imitations by the false on the cow's udder and nipples; and all is called
> Cow Pox whether on the cow or communicated to the human animal.

Jenner's defence, born of careful observation and analysis, was that
only genuine cowpox, not other similar eruptions on cows, was effective;[26] no pustule must be excessively raided; and (as he had learnt from
Sutton in smallpox inoculations) the vaccinating fluid had to come
from the pustules in the early stage of their development, whether this
was from the cow or from humans. He added that to avoid complications, it was important that vaccination should be carried out only on
healthy children, free of eczema or any other skin problem. Very occasionally a properly vaccinated subject might prove vulnerable to smallpox, he admitted, just as very occasionally someone might get smallpox
after inoculation, or even get natural smallpox twice, but 'duly and
efficiently performed, [vaccination] will protect the constitution from
subsequent attacks of small-pox, as much as that disease itself will.
I never expected that it would do more, and it will not, I believe, do
less.'[27]

The next month saw Woodville's report on the vaccinations he and
Pearson had carried out. Although Woodville remained convinced
that vaccination was a great improvement on variolation, his results
seemed to show that out of 500 cases vaccinated 300 had numerous
pustules, and in at least two cases this led to the infection of others.[28]
The trouble was that the Smallpox Hospital was not an ideal place
for vaccination, for patients were exposed to smallpox infections, the
material supposed to be taken exclusively from cowpox pustules was
(it was later confirmed) sometimes contaminated with the contents of
smallpox pustules, and Woodville had occasionally used matter from
the pustules of the general eruptions for further insertions. The following year Woodville himself confessed he was baffled by the effect of the

'variolous atmosphere' of the hospital.[29] All the same, it was some time before habits improved. Joseph Adams, when he succeeded Wood-ville, still had cowpox and smallpox preparations together on his table, so that parents could choose which they wanted – in 1806 the sixty or seventy who came each day were about equally divided.[30] For the next two years out-patients could still choose between variolation and vac-cination, and in-patients had the choice until 1822.[31] Luckily the 200 threads with vaccine material that Pearson had sent out in March 1799 probably came directly from Gray's Inn Lane or Marylebone Fields, not from Woodville's hospital. But there were occasions, harmful to the reputation of vaccination, when infected vaccine material was sent out from the Smallpox Hospital. In a notorious incident at Petworth, in Sussex, all fourteen people vaccinated had an eruptive reaction, as if inoculated with smallpox; they recovered, but a woman nursing them caught smallpox and died.

Hating the smoke and the jealous medical politics of London (a hatred that was confirmed by an abortive attempt to settle in London in 1803), Jenner returned to Berkeley, now a country doctor with a mis-sion, passionate about spreading the news about cowpox. But matters were growing out of his control. He had only just got back when, with the worthy aim of vaccinating the poor free of charge as well as distrib-uting vaccine round the country, Pearson set up the Institution for the Inoculation of the Vaccine-Pock. Jenner, to his fury, was offered the insulting position of 'extra consulting physician'. It was the first move in a distressing series of quarrels that rumbled on through the rival establishments of the Royal Jennerian Society (1802) and the National Vaccine Establishment (1808).

—·—

There was plenty of 'cavil and controversy', together with prejudice over the whole idea of vaccination. A London debating society dis-cussed 'Which of these two propositions affords the greatest proof of human credulity, that light can be obtained from gas, or that protec-tion from smallpox can come from vaccination?'[32] One of the chief

19. 'The Cow-Pock – or – the Wonderful Effects of the New Inoculation.'
*James Gillray's 1802 cartoon mocking the more exotic fears of what vaccination
might lead to.*

antagonists was Benjamin Moseley, a physician with various cranky
ideas (he believed the phases of the moon influenced haemorrhage from
the lungs) who had worked in the West Indies and had written about
tropical diseases. In 1800 he published a volume on sugar, cowpox, the
yaws, African witchcraft, the plague, yellow fever, hospitals, goitre and
prisons.[33] Hidden in this hotch potch is a swipe at vaccination: 'Can
any person say what may be the consequence of introducing the *Lues
Bovilla* [syphilis of oxen], *a bestial* humour – into the human frame …
the doctrine of engrafting distempers is not yet comprehended by the
wisest men.' It was the start of a hysterical campaign, with Dr William
Rowley claiming that vaccination had produced an 'oxfaced boy' and
descending to placarding London urinals with 'proofs against cowpox'.[34] Gillray's cartoons ridiculed the hysteria by showing vaccinated
patients sprouting various cattle-like features. John Birch, surgeon at

St Thomas's Hospital, announced that vaccination was useless and, by replacing inoculation, would allow smallpox to 'recur in all the terrors with which it was first surrounded';[35] he also put forward the harsh argument that smallpox was 'a merciful provision on the part of Providence to lessen the burden of a poor man's family'.[36] His tomb was inscribed:

> But the practice of Cow Poxing
> Which first became general in his day,
> Undaunted by the overwhelming influence of power and prejudice
> And the voice of nations
> He uniformly, and until death, perseveringly opposed;
> Conscientiously believing it to be a public infatuation,
> Fraught with perils of the most mischievous consequences to mankind.

The cold voice of Malthus had already been heard. His *Essay on Population*, talked of smallpox as 'one of the channels, and a very broad one, which nature has opened for the last thousand years to keep down the population to the level of the means of subsistence'. In 1826, appalled at the progress of vaccination, he went further: 'We should reprobate specific remedies for ravaging diseases, and those benevolent, but much mistaken men, who have thought they were doing a service to mankind by projecting schemes for the total extirpation of particular disorders.'[37]

But there were plenty of supporters for Jenner. One of the most prolific was his friend John Ring, who 'published one thousand and forty chaotic pages in defence of the new practice' in 1801.[38] Some had changed their minds reluctantly – in 1802 a Mr Thomas of Daventry wrote to *The Medical and Physical Journal* that 'to my very great disappointment, I found the Vaccine-pock so safe and mild a disease … that I became a convert … and in a very short time [was] compelled to shut up my Inoculating Houses'.[39] John Lettsom was a notable convert from inoculation. It was Lettsom, once the passionate defender of the Society for the Inoculation of the Poor, who first sent a copy of Jenner's

'Inquiry' to America, to his fellow Quaker Benjamin Waterhouse at Harvard, and wrote a helpful pamphlet 'Observations on the Cow-Pock' (1801). In London, vaccination was one of the many good causes taken up by the evangelical preacher Rowland Hill, who published 'Cow-pock Inoculation Vindicated and Recommended from matters of Fact', and himself vaccinated thousands. An impressive early report of the success of mass vaccination came from a village in Westmoreland in 1800 where, as soon as a case of smallpox appeared, the local doctor Robert Thornton was ordered by the landowner to vaccinate all 400 inhabitants; this stopped the smallpox from spreading, and in the next few weeks Thornton carried out about 1,000 more vaccinations in the neighbourhood.

In 1800 the Medical Faculty of Oxford gave Jenner a testimonial: 'We, whose names are undersigned, are fully satisfied upon the conviction of our own observation, that the Cow Pox is not only an infinitely milder disease than the Small Pox, but has the advantage of not being contagious, and is an effectual remedy against the Small Pox.' The Earl of Berkeley raised a subscription to give Jenner an award of some piece of plate. 'Have you thought of an appropriate device?' Jenner wrote cheerfully to his friend Henry Hicks. 'What think you of the *cow jumping over the moon*? Is it not enough to make the animal jump for joy?'[40]

More tributes flowed in to keep Jenner cheerful in spite of the worries and the opposition. A petition was presented to Parliament in 1802 in Jenner's name asking for 'such remuneration as to their wisdom shall seem meet', because (as had also been the case with his 'tartar emetic' earlier) he had not selfishly kept his ideas in any way secret, but had been so concerned to 'promote the safety and welfare of his countrymen and of mankind in general' that he had actually lost money by not having time for his usual practice.* Witnesses said that if Jenner

---

*Such awards for distinguished service did sometimes take place, but were rare – Harrison had been given nearly £8,000 in 1773 for his chronometer, and in the same year as Jenner's petition Dr Carmichael Smyth was granted £5,000 for his discovery of the use of 'nitrous fumigation' against infection.

had chosen a different approach he could have gained an income of
£10,000 rising to £20,000; as it was, his success meant that his mail
often cost him over £1 a day, and took so much time that he called
himself 'Vaccine Clerk to the World'. Pearson now took on the role
of detractor, pushing Jesty forward as the originator of vaccination,
but Parliament decided that 'on the whole, vaccination was considered
to be the greatest discovery ever made in medicine, and Dr Jenner the
sole discoverer'. The suggestion of a grant of £20,000 was narrowly
defeated, and Jenner was given £10,000. (This was just at the time
when Jane Austen's wealthy Mr Darcy had £10,000 a year, and Mr
Bingley only half as much.) 'Your friends made a bad business of your
application to Parliament', Lord Sherborne wrote, 'as I think you the
greatest patriot that ever existed; and you ought to have had at least
£50,000.'[41] Five years later, with the support of the College of Physicians, this rather mean treatment was improved by a further award of
£20,000. Grateful subscriptions came from the British in India too,
from Calcutta, Madras and Bombay.

There were many other official honours. In 1802 John Quincy
Adams invited Jenner to be a member of the American Academy of
Arts and Sciences, and the next year Harvard gave him an honorary
LL.D. He was made a Freeman of the cities of London, Dublin and
Edinburgh, a foreign associate of the National Institute of France, and
given an honorary Doctorate of Medicine by Oxford – the first to be
awarded for seventy years. Even so, the College of Physicians, sticking
to its rules, refused to admit him as a Fellow without an examination in
Greek and Latin: 'In my youth I went through the ordinary course of
a classical education', Jenner wrote, 'obtained a tolerable proficiency in
the Latin language, and got a decent smattering of the Greek; but the
greater part of it has long since transmigrated into heads better suited
for its cultivation. At my time of life to set about brushing up would be
irksome to me beyond measure: I would not do it for a diadem.'[42]

Far more important than all these, to Jenner, was the knowledge of
the enormous change he was bringing into people's lives. 'A few years
ago I was in the habit of burying two or three children every evening

in the spring and autumnal seasons, who had died in small-pox', the Revd Mr Finch wrote to him from St Helen's in Lancashire, 'but now this disease has entirely ceased to call a single victim to the grave. Why? I have inoculated for the cow-pox upwards of 3000 persons, and the small-pox is no longer in existence here.'[43] In the last decade before vaccination (1791–1800), the London Bills of Mortality showed a total of 18,447 deaths from smallpox; in the last decade of Jenner's life (1811–20), this had dropped to 7,858. It was a success story that was repeated around the world.

9

The world-wide spread

Away from the squabbles in England, Jenner's reputation and the use of vaccination spread amazingly fast in Europe and beyond. 'From Manilla and the Philippine Islands they send me an account of 230,000 successful cases', Jenner wrote proudly to Richard Dunning in 1806. 'From Canton I have a most curious Production; a pamphlet on vaccination in the Chinese language. Little did I think, my friend, that Heaven had in store for me such abundant happiness.'

Pearson had sent vaccine around Europe as well as around Britain. The *Bibliothèque britannique*, which was founded in 1796 in Geneva to keep a war-torn continent in touch with British liberal ideas and intellectual news, had as its medical editor the inoculator Louis Odier. He was so enthusiastic that, as well as publishing extracts from the 'Inquiry' only three months after it had appeared in England, he also wrote a pamphlet on vaccination, for clergymen to give to parents when they brought their children to be baptised. In England others had similar ideas – Jenner's biographer, John Baron, mentions letters to Jenner from the Revd Dr Booker of Dudley, saying that he too handed out pamphlets, and from Erasmus Darwin (who was himself badly pock-marked) suggesting that vaccination should be done at the same time as christening. There are even records of some clergymen in England, Switzerland and Germany carrying out vaccinations themselves.

The first successful vaccinations in Vienna were done with material sent by Pearson. They were witnessed by Dr Jean de Carro, a refugee from the revolutionary atmosphere of Geneva, who had studied

in Edinburgh and had a great reputation among ambassadorial circles and the English colony in Vienna. He became the local apostle of
vaccination, seeing his way to fame and fortune. Letters over the next
five years to his friend Alexandre Marcet tell the story. Marcet had the
same background of Geneva and Edinburgh, but he had remained in
London after his studies, and was used as the source of information.[1]
Mixed with accounts of triumphs of the Austrian army, de Carro sent
requests for anything that had been written about vaccination, and
questions: had it been shown that vaccination was as effective when
it had passed through several individuals? Would raising the skin by
a vesicatory (a blistering ointment) have the same effect as an incision?
Couldn't the problems of eruptions after vaccinations at the London
Smallpox Hospital be the result of contamination by smallpox? Was
it true that the Duke of York had ordered vaccination for everyone in
his regiments who had not had smallpox? De Carro vaccinated his
own two children, confirmed the success by smallpox inoculation,
arranged with Marcet for further supplies to be sent to Vienna by diplomatic bag and, like Henry Cline, declared vaccination to be 'une des
plus belles découvertes de la Médecine'. In the summer of 1803, remembering how much opposition he had encountered at first, he was proud
to tell Marcet that he had not seen a single case of smallpox in Vienna
for two and a half years, nor had an oculist seen any cases of 'ophthalmie
variolique' – and he used to see more than fifty cases in his practice every
year. This happy state of things did not last, for three years later we find
de Carro writing about poor children dying of smallpox in the Vienna
suburbs.[2]

It was, indirectly, de Carro who introduced vaccination to India.
Meeting Mr and Mrs Nisbet, who were on their way to visit their
daughter Lady Elgin – wife of the marble-collecting British ambassador to Constantinople – de Carro convinced them of the importance
of vaccination.[3] He luckily had a stock of vaccine from cowpox that
had recently been discovered by the vaccinator Luigi Sacco in a herd in
Lombardy, so he was able to supply them with vaccine for Lord Elgin's
young son, whose pustule in turn was a link in the chain that in 1801

supplied vaccine throughout Turkey and Greece – introducing vaccination to the countries where another British ambassador's wife had
encountered inoculation more than eighty years earlier.

Elgin arranged for a further supply of vaccine from de Carro to be
sent to Constantinople and, Lady Elgin wrote to her mother,

> *it is quite astonishing how much it has taken. There have already been
> seventy people and children inoculated, and the people at Belgrade [a
> summer resort near Constantinople] have sent and begged to be vax-
> ined. Elgin has had many letters from Smyrna intreating him to send
> some vaxine there; hundreds and hundreds of children are dying every
> day of the small pox at Smyrna, and at Pera too it is very fatal. I think
> we shall completely establish vaxine in this country.*[4]

At that time there was no reliable way to preserve vaccine for a long
and hot journey, so the problem was tackled by a series of arm-to-
arm vaccinations. Matter from de Carro's consignment was sent on
to Baghdad, and then used to vaccinate crew and passengers on a
ship bound from Basra to Bombay; a three-year-old child was vaccinated in Bombay in 1802 and, the Bombay Medical Board reported,
'from her alone the whole of the matter that is about to be sent all
over India was at first derived'. It was sent on to Ceylon the same
year with considerable success – within three years, 33,000 Ceylon-
ese had been vaccinated, and by 1821 endemic smallpox there had
almost been defeated.[5]

Native opposition to vaccination in India was partly overcome by
what has been called a 'pious fraud':

> *Mr Ellis, of Madras, composed a short Sanscrit poem on the subject
> of vaccination. This poem was inscribed on old paper, and was said to
> have been found, that the impression of its antiquity might influence the
> minds of the Brahmins, whilst stress was laid upon the fact that the
> benefit was to be derived from their sacred cow.*[6]

An extra push was given by Richard Wellesley, Governor-General and brother of the Duke of Wellington, who offered the Brahmins double the profits they had been earning as inoculators if they took up vaccination.

Jenner had thought of sending vaccine to India on a ship with twenty volunteers, one of whom would be vaccinated before sailing, so that each in turn could be vaccinated on the way; it was to have been financed by subscription, and he had himself offered to put up 1,000 guineas.[7] Elgin's success meant that this was never needed, but a voyage with arm-to-arm transfer on board ship was organised by a French physician to spread vaccination from India to the French colonies of Île de France and Réunion (near Mauritius); from there it was sent similarly to Batavia (now Jakarta) in Java.

A grand version of this system was financed by King Charles IV of Spain. Charles IV has been given a bad press as a weak and stupid king, his enthusiasm for vaccination hardly mentioned except by medical historians. But he had set up the Royal College of Surgery and Medicine in Madrid, he had read Jenner, and he determined 'to ameliorate the havoc wrought by the frequent smallpox epidemics in his dominions of the Indies'.[8] So, moved also by the thought that it would be economically sound to halt the depopulation of the Spanish colonies, in 1803 he commissioned the royal physician Dr Francisco Xavier Balmis to undertake a great expedition. Balmis loaded his ship with goods to trade, two physicians, four assistants, three male nurses and twenty-two foundlings aged between three and nine (one of them just vaccinated), together with the matron from the foundling hospital at Santiago; his mission was to take cowpox vaccine to all the Spanish colonies, vaccinating the boys in pairs every ninth or tenth day. He called in at the Canaries, where he set up a vaccination clinic, and at Puerto Rico, where he was dismayed to find vaccination had already been introduced by threads brought from England to a nearby island. By the time he arrived in Caracas, only one of his boys still had a viable pustule, but that was enough to start vaccination on the South American

continent. The twenty-two little Spanish foundlings were then settled in Mexico, maintained and educated at the expense of the Spanish
treasury, and eventually adopted.

In Venezuela the expedition divided, recruiting more children and
spreading vaccination around the continent. One group, surviving a
shipwreck on the coast of Colombia, crossed the isthmus of Panama
and carried out 50,000 vaccinations along the Peruvian coast before
going on to Buenos Aires, Chile and the Philippines on the way back
to Spain. The other group, led by Balmis, went to Cuba, Yucatán and
Mexico. He took twenty-six children on board for the journey on to
the Philippines; they were taken back to their homes and to a promised
reward of education two years later. Before returning to Spain, Balmis
took three boys from Manila to start a chain of vaccinations in Macao,
and then one Chinese boy across to Canton; finally, he introduced vaccination to St Helena.[9] 'What a glorious Interprize', Jenner wrote to
the printer Richard Phillips. 'I have made Peace with Spain [then at
war with Britain] and quite adore her philanthropic monarch.'[10]

It was glorious and it was spectacular, but it was not unique. The
Portuguese too, anxious to introduce vaccination to their Latin American colonies, tackled the problem in a similar way but from the other
direction. In 1804 they recruited slave children in Bahia (Brazil), and
sent them to Lisbon so that they could be vaccinated arm-to-arm on the
return voyage.[11] The vaccine set up a successful chain in Montevideo
(Uruguay) as well as Bahia.

The uprooted children on the Balmis expedition seem to have been
well cared for, but the whole affair highlights the way foundlings were
so often looked on as useful material for experiments – like the 'half a
dozen of the charity-children belonging to St James's parish' inoculated before the Princess of Wales would risk it on her own children.
It was an accepted practice throughout Europe, reflecting not only a
disturbing ethical code, but also the appalling numbers of abandoned
children – many illegitimate, and many simply left by mothers too
poor to cope. It was said that in Paris at the end of the eighteenth century, about 6,000 children were orphaned or abandoned each year out

of a total population of 600,000.[12] Italy, France and Russia had long had foundling hospitals, and Thomas Coram, moved by the sight of infants abandoned in the streets, set up the London Foundling Hospital in 1739. The Empress Dowager of Russia, anxious to stop the drain of manpower from smallpox, introduced vaccination to Russia with a demonstration on an orphan – the child was then named 'Vaccinoff' and compensated with education and a pension. Luigi Biagini, in Tuscany, wrote that the foundlings of Pistoia were useful in vaccination experiments, and shamelessly admitted that he took more risks with them than with his private patients. De Carro envied him, telling him that in his own rather high-class practice he had not the same opportunities.[13] Yves-Marie Bercé comments in *Le chaudron et la lancette* that this was not because Biagini and de Carro considered some lives more valuable than others, but because they were scared of the bad publicity for vaccination if anything went wrong in a prominent family.

———

Apart from an eighteen-month gap from the end of 1801, Europe was almost constantly at war from 1793 to 1815. But, as Jenner wrote, 'The Sciences are never at war.'[14] Jenner's 'Inquiry' was translated into German, French, Italian, Dutch, Spanish and Portuguese, and the French gave a special passport to William Woodville in June 1800 so that he could bring vaccine from the London Smallpox Hospital to Paris and instruct the French doctors who had not been having much success. They had vaccinated an unfortunate batch of orphans from the Hôpital de la Pitié, using material sent by Pearson, which may have been spoilt by the hot weather, for it did not give proper reactions. Woodville arrived with his new stock, accompanied by Dr Antoine Aubert, who had crossed to England to see the Smallpox Hospital at work, and by Thomas Nowell, an English doctor with a practice in Boulogne. After tiresome delays over permits, Woodville reached Paris, but found his vaccine was no longer effective. Luckily Nowell had successfully vaccinated three children in Boulogne on the way, and that 'matière de Boulogne' became the source of vaccine that spread around France and neighbouring countries.[15] Woodville, 'a solemn taciturn Englishman'[16] overwhelmed by the enthusiasm of the French, was able to use it in Paris – after, of course, he had tried it on yet more orphans. The statue of Jenner in Boulogne honours both Woodville and Nowell in its inscription.

The Napoleonic wars actually helped the introduction of vaccination to the Mediterranean region. In 1800 the Duke of York, concerned that the navy and the army overseas should be protected as well as the army at home, sent 'two eccentric doctors',[17] Joseph Marshall and John Walker, to work in Gibraltar, Minorca and Malta. A sailor on board their ship was vaccinated, and his arm started the chain for the whole project. After clearing Malta of smallpox, Walker went on with the British expedition to Egypt, contributing to the army's success against the French troops there by making sure the British were safe from small-pox. Marshall went to Palermo, where 8,000 had died from small-pox the previous summer, and set up a vaccination centre in a Jesuit seminary where he treated the poor without charge twice a week. He described it to Jenner:

> *It was not unusual to see in the mornings of the public inoculation [vac-*
> *cination] at the Hospital a procession of men, women and children,*
> *conducted through the streets by a priest carrying a cross, come to be inocu-*
> *lated. By these popular means it met not with opposition, and the common*
> *people expressed themselves certain that it was a blessing sent from*
> *Heaven, though discovered by one heretic and practised by another.*[18]

Having trained other vaccinators and having carried out, with their help, more than 10,000 vaccinations in Sicily, Marshall moved on to Naples, where he set up an Institution for Jennerian Vaccination. The brief interval of peace (1802–3) allowed him to travel overland to Paris through Rome, Leghorn, Genoa and Turin, spreading the benefits of vaccination as he went. It was badly needed; in the slums round the port in Genoa, for example, beggars would parade their pustule-covered infants in a bid for charity.

Vaccination had already arrived in parts of north Italy, where Luigi Sacco (the man who had found the Lombardy cowpox) had become Director of Vaccination for Napoleon's Cisalpine Republic in 1801. The migrant workers of Italy spread infections, and the many isolated settlements were vulnerable, but Sacco boasted that within three years he had got rid of smallpox in Lombardy. 'I flatter myself', he wrote in 1808, 'that in Italy I have been the means of promoting vaccination in a degree which no other kingdom of the same population has equalled.'[19] He had sent Jenner some cowpox lymph in 1802; ten years later Jenner sent him some in return, through the Milan agent of the London art dealer Colnaghi. In his 1813 treatise on vaccination, Sacco claimed that he himself had done more than half a million vaccinations, while a further 900,000 had been done by his colleagues – a conscientious doctor would even travel to remote alpine villages taking a recently vaccinated boy to supply the cowpox. When smallpox struck Rome in 1814, vaccination was endorsed by Pope Pius VII. 'Almost all the new-born children are vaccinated,' Sacco reported to Baron in 1824, 'so that we have now no fear of the small-pox. It is occasionally imported from the neighbouring states of Parma, Piedmont, &c. Such occurrences never fail to prove the efficacy of the preservative, for the disease never becomes epidemic.'[20]

Napoleon and Jenner made use of each other. '*Ah, Jenner, je ne puis rien refuser à Jenner,*' Napoleon famously said in response to an appeal from Jenner for the release of two Englishmen who had been stranded in France and interned when war started again in 1803. Napoleon was quick to realise the value of vaccination. He had been a consul (one of the three governing France) at the time of Woodville's success in Paris, and had seen the establishment of the Comité Central de Vaccine which did so much to spread the practice around France. Free vaccination centres had been set up, health officers were sent round the country to teach and to vaccinate. 'Glory and recognition to the inventor and propagators of the process with whose assistance we are saving the human race from the scourge that was decimating it', proclaimed one local council.[21] In 1802, on his arrival in Paris, Joseph Marshall joined Edward Jenner's nephew George at an official dinner in honour of his uncle. When Napoleon was Emperor, he ordered that FR100,000 should be used to encourage and propagate vaccination. A vaccination medal was designed, with Napoleon on the front and a small cow and a lancet flanking Venus and Aesculapius (the Greek god of medicine) on the back; commemorative china appeared too, honouring vacci-nation and Jenner. In the year of the battle of Waterloo, the French Academy announced the subject of its poetry competition would be Jenner and vaccination. (While Napoleon was promoting vaccination for the French nation, Nelson was writing to persuade Lady Hamil-ton to arrange it for their daughter Horatia; Lady Hamilton was not persuaded, and the three-year-old, as she herself said later, 'took the smallpox very severely and was not expected to live'.[22])

British recruits who were not already immune were vaccinated, which made, as Jenner wrote, a contrast to the 'incessant losses by small-pox among the troops in former wars'. It was not until 1805 that Napoleon ordered the same for his soldiers; but in spite of his sound proposal that two soldiers from each battalion should be vaccinated to start a chain for others, there was no compulsion and fewer than half the vulnerable men were vaccinated. New regulations in 1811 were more effective, but peace and the downfall of Napoleon came four years later, and the rules were relaxed. The break-up of the Empire, bringing general instabil-ity and also a reaction to ideas associated with the Napoleonic regime, brought a temporary set-back to the progress of vaccination.

Figures from all round Europe showed the dramatic fall in smallpox deaths when vaccination was introduced. In Sweden, annual deaths per million population, which had reached nearly 2,000 in the decade 1792–1801, fell to an annual average of 133 in the decade 1812–21. In the same periods, the total number of smallpox deaths in Berlin fell from 5,000 to 555 – helped by the Royal Inoculation [Vaccination] Institute there, which gave a medal to poor children who came back seven days after vaccination to have their arms checked. The King of

20. Smallpox epidemic in Cape Town

Prussia, in 1799, had been the first ruler to have his own family vac‑
cinated. In Copenhagen, where smallpox deaths had averaged 373 a
year before vaccination, there were no deaths at all from smallpox for
twelve years from 1811.[23]

News of vaccination reached Cape Town in 1800 with a series of
articles printed in English and Dutch in the *Government Gazette*, but
it was another three years before vaccine arrived – on a Portuguese
ship where it had been maintained arm‑to‑arm on slaves. For many
years after that, vaccine was supplied by South America, as it was
difficult to keep up an arm‑to‑arm supply in a country with only
periodical, though disastrous, epidemics. An enthusiastic Vaccine
Institute was started, and in spite of some resistance to the idea of

regulation, district surgeons tried to carry out a systematic plan of vaccination. Unprotected tribes continued to suffer – an epidemic in 1831 killed nearly 90 per cent of the Griquas.[24] In 1840 an epidemic spreading particularly badly in the crowded poor districts in Cape Town killed around 2,500. That time the infection probably arrived with a slave ship, but this hazard decreased as the traders came to realise that vaccinating the slaves was good for business. There was very little vaccination in the rest of Africa, apart from Egypt, before 1900.

Vaccination was tried in North America very shortly after Jenner's 'Inquiry' was published. A friend from Jenner's schooldays in Cirencester, John Clinch, had become a missionary doctor in Trinity, Newfoundland, and Jenner sent a packet of dried vaccine to him in 1798. A local paper gave an account of what happened:

> *Although small-pox was prevalent, Dr Clinch could not persuade anyone to try the new method, and eventually he applied the vaccine to his nephew, a boy of about seventeen years of age, who submitted to treatment by no means willingly. The application having proved effective, such was Dr Clinch's confidence in Jenner, that he placed the boy in bed with one of the worst cases of small-pox at that time under his attention. To the surprise of everyone he did not contract the disease, and immediately there were insistent demands from everyone for the treatment.[25]*

Clinch was able to respond to the demand with a chain of around 700 more vaccinations, the first in North America. Canadian vaccination seems to have stopped at that stage, until further supplies of vaccine came from England in 1802. A Bureau de Vaccine was established in Quebec in 1821.

It was Benjamin Waterhouse in Boston, inspired by the copy of the 'Inquiry' sent to him by Lettsom, who set off the major American campaign. He got his vaccine from John Haygarth, the man who had

21. Benjamin Waterhouse

once hoped that smallpox could be defeated by a regulated programme of inoculation. Waterhouse started by vaccinating his son and six members of his household, and then, in the manner of Jenner, testing them with smallpox inoculations. But his next move was more reminiscent of the mercenary approach of Daniel Sutton – he tried to keep a monopoly of North American vaccine, charging other practitioners a fee or a quarter of the profits. It was a disastrous policy as well as a selfish one, for it led to a bootleg market in bits of clothing infected with discharges from pustules which might be contaminated, or no longer active, or even derived from smallpox not cowpox. At this stage material from a pustule on the arm of a sailor from London, thought to have been the result of vaccination, was used by a physician in Marblehead to vaccinate his daughter and, from her, several others; but the sailor had been infected with smallpox, and sixty-eight died. Fortunately several other doctors started to get vaccine from England, restoring the reputation of vaccination and breaking Waterhouse's monopoly.

Waterhouse, accepting defeat, freely distributed his next supply, and from then on promoted vaccination in a properly public-spirited way. He persuaded the Board of Health of Boston to vaccinate nineteen volunteers, then inoculate them with smallpox and publicise the successful result. 'This decisive experiment', he announced, 'has fixed for ever the practice of the new inoculation in Massachusetts.'[26] The first American community to arrange vaccination for its citizens was Milton, Massachusetts – in 1809 the selectmen hired a doctor to vaccinate for 25 cents per person (instead of the usual $5). Over 300 citizens responded (more than a quarter of the total population of the town), leaving only twenty unprotected by either vaccination or previous infection. Benjamin Russell, the president of the Boston Board of Health, proclaimed that 'The example set in *Milton* of a general inoculation for the *Kine* [or Cow] *Pock*, has done that town honor; and we hope it will be emulated in every town, not only of this, but of every State of the Union.'[27] He did his best to make this happen, promoting bills in the Senate. The law that emerged encouraged vaccination rather than ordered it, which limited the effect, but in the war of 1812 the American army was almost clear of smallpox, and from 1816 to 1824 there were no deaths from smallpox in Boston.

Waterhouse's next influential move was to write to the new President, Thomas Jefferson, sending him some vaccine. Jefferson was enthusiastic. He arranged for the vaccination of his own household (including the slaves), and his sons-in-law and his neighbours did the same – about 200 people altogether. He designed a water-cooled container which held the vaccine in an inner phial, saw that vaccine was distributed to Washington and Philadelphia, and with both diplomacy and humanity gave some to a group of Indians. 'The Great Spirit,' he told them, had 'made a donation to the enlightened white men; first to one in England, and from him to one in Boston, of the means to prevent them from ever having smallpox'; the Indians were then vaccinated, and taught how to vaccinate.[28] Jefferson also told the explorers Lewis and Clark to take cowpox vaccine with them and promote its use among the Indians. In 1807 Jenner contributed to the good work by sending a book explaining vaccination to the Chief of the Five Nations, assembled at Fort George in Upper Canada, who gratefully sent back a letter with a belt and a string of wampum.

Jenner had a grateful letter from Jefferson too. 'You have erased from the calendar of human afflictions one of its greatest', he wrote. 'Yours is the comfortable reflection that mankind can never forget that you have lived; future nations will know by history only that the loathsome small-pox has existed, and by you has been extirpated.'[29]

More news reached Jenner from all around the world. The pamphlet on vaccination written in Chinese that was sent to him in 1806 had been composed by a surgeon of the East India Company, translated into Chinese, and distributed at the expense of the Company. It was followed by others written by the Chinese themselves. In a report to the National Vaccine Establishment of Canton ten years later the surgeon Alexander Pearson told how vaccination had first come to China. Balmis, he claimed, had not been the first: '... before his arrival in China, it had been quite extensively conducted by the Portuguese practitioners at Macao, as well as by myself among the inhabitants there and the Chinese'.[30] Pearson added that, diplomatically, he had first treated only the poor, and at his own expense, but soon the method

> sprang into favour amongst the Chinese, who though very conservative in their feelings, when once convinced of the benefit of any new method, take it up very readily and great numbers were brought to be operated on during the period of the raging smallpox in the course of the winter and spring months of 1805–1806.[31]

There were problems, once an epidemic was over, in persuading enough parents to allow their children to be vaccinated to keep the arm-to-arm method going, but in 1815 a free dispensary, run by Chinese vaccinators, was set up in Canton, and by the time Pearson left China seventeen years later the practice was well established. Japan at this time, firmly closed to outside influences, carried on the Chinese method of inoculation and still had to wait for vaccination.

News from Java came in 1816 when Jenner was visited by Sir Stam-ford Raffles, back from a stint as Lieutenant-Governor. Raffles had tried to tackle the problem of smallpox there, organising the training of native assistants for vaccination and paying them with allotments of land. A great collector with a passion for wild animals – he became the founder of the London Zoo – Raffles shared many interests with Jenner, hearing his memories of John Hunter and the strange Earls Court menagerie.[32]

Jenner died in 1823, twenty-five years after the publication of his experiments, having seen vaccination spread to four continents. Even as early as 1801 he had been able to write:

> *A hundred thousand persons, upon the smallest computation, have been inoculated [vaccinated] in these realms. The numbers who have par-taken of its benefits throughout Europe and other parts of the Globe are incalculable; and it now becomes too manifest to admit of contradic-tion that the annihilation of the smallpox, the most dreadful scourge of the human species, must be the final result of this practice.*[33]

10

Confusion and compulsion

The first twenty years after the 'Inquiry' saw millions vaccinated world-wide, and a rapid drop in the mortality from smallpox. But for Jenner himself, the pride and happiness brought by his success were always mixed with frustration. Things were out of his control and not going as well as they could and should. Unforeseen problems – particularly cases of smallpox in supposedly vaccinated subjects – were reported, and exploited by his opposers. There were also new outbreaks because vaccination was not yet universal and variolation was still in use.

Encouragingly though, the College of Physicians had reported in 1807:

> The security derived from vaccination against the small-pox, if not absolutely perfect, is as nearly so as can perhaps be expected from any human discovery; for among several hundred thousand cases, with the results of which the College have been made acquainted, the number of alleged failures has been surprisingly small ... not nearly so many failures in a given number of vaccinated persons as there are deaths in an equal number of persons inoculated for the small-pox.

If smallpox did occur in a vaccinated subject, they added, it was mild. And, unlike variolation, vaccination 'communicates no casual infection, and, while it is a protection to the individual, it is not prejudicial to the public'. They warned against the persistent use of variolation, and recommended that vaccination should be offered 'to the poorer classes without expense'.[1]

An epidemic in Norwich in 1818–19 showed how easily disaster could hit a town which had been free of smallpox for five years and where only 8,000 had been vaccinated out of a population of 50,000.[2] A girl who had come with her parents from York arrived ill and infect-ed several others, two of whom died. A druggist then variolated three children 'thereby helping to keep up the contagion'. It spread round the city: more than 3,000 caught smallpox and 530 died. But in all

> *not more than 1 in 20 vaccinated persons will be found to be in any way affected by the most intimate exposure to small-pox; or less than 1 in 50 will have the disease in a form answering even to the generally received description of modified small-pox.*

To avoid the disease spreading in the nearby borough of Thetford, as soon as there was one case:

> *The parish officers visited every house, made a list of all those liable to the contagion, and threatened to expose any individual who should refuse vaccination, or submit clandestinely to variolous inoculation. The list thus made was delivered to the two surgeons, and the bellman was employed to announce the hour on the following morning at which all those requiring it might be vaccinated at the churches of their respective parishes. These prompt means were apparently successful. About 200 were vaccinated, most of them in the course of two days, and small-pox extended only to eight or ten persons, all of whom survived.[3]*

Edinburgh too suffered an epidemic in 1818, and here there were reput-edly 1,500 cases of smallpox in vaccinated individuals, three of whom died. Jenner blamed this on incompetent vaccination; with good vac-cinators such as Luigi Sacco, he said, things did not go wrong. J. F. Marson, surgeon at the London Smallpox Hospital for over twenty years, supported Jenner's point, writing that there were many 'bun-gling operators' and that, of more than 3,000 supposedly vaccinated persons whom he had seen suffering from smallpox, only 268 presented what he considered the marks of thorough vaccination;[4] and all too often patients failed to return to check that the vaccination had 'taken'. But the Edinburgh epidemic also illustrates the general confusion over facts and figures. Dr Thomson, who wrote an account of the outbreak, admitted that there was epidemic chickenpox at the time, and that considering it to be the same disease he had included all those cases in the total.[5] He was not the only one – in spite of William Heberden's clear paper a century earlier, Dr Hamernik of Prague, in an extraordinary attack on vaccination for John Simon's 1857 report to Parliament, said firmly that chickenpox was a form of smallpox. The contribution to the report from the surgeons of Vienna also confused matters by talking of 'three species of small-pox' – the dangerous form (variola vera), modified smallpox (varioloid), and chickenpox.[6]

It was not only 'spurious cowpox', Jenner thought, that could spoil vaccinations. He wrote an article in 1804 on the effects of 'cutaneous eruptions', and reprinted and circulated it as a pamphlet, 'On the Varieties and Modifications of the Vaccine Pustule, Occasioned by an Herpetic State of the Skin'.[7] 'One more word on herpes', he wrote to Dunning. 'Seeing how frequently vaccine disease becomes entangled with it, my thoughts have been pretty much bent upon it ... A single [herpetic] vesicle is capable of deranging the action of the vaccine pustule.'[8] Jenner, struggling to work out his theories without the help of modern knowledge of infections, had come to the right conclusion for the wrong reason. (Seventy years earlier, Hans Sloane had also realised that it was dangerous 'to inoculate [variolate] such, as have any breakings out on their faces, soon after the measles, or any other occasion, whereby the small-pox were likely to be invited, and come in the face in greater number'.[9]) Dixon has warned that, unless there is an immediate risk of smallpox, it is wise to postpone vaccination where there is an infection of the skin, not because the infection interferes with the effect of the vaccine, but because the infected areas may become contaminated with the vaccine and produce further pustules.[10]

A problem which became increasingly clear over the years, but which was never recognised by Jenner, was that vaccination did not give permanent protection. His friend Dr Percival had raised the point as soon as he read the 'Inquiry', saying in his letter of thanks for the pamphlet that 'a larger induction is yet necessary to evince that the virus of the *variolae vaccinae* renders the person who has been affected with it secure during the whole of life, from the infection of the small pox',[11] as Jenner had rashly asserted. William Goldson, an antivaccinationist from Portsmouth, printed his doubts about permanence in a pamphlet in 1804;[12] John Ring loyally but misguidedly countered with 'An Answer to Mr Goldson, proving that Vaccination is a Permanent Security against the Small-Pox'.[13] Ring became Jenner's great correspondent and supporter; four years later he went 'at the head of a deputation to investigate some supposed failures of vaccination. The anti-vaccinationists were put to shame, but party feeling ran so high that the deputies carried pistols to defend themselves in case of need.'[14]

The sparring over permanence began an argument that was to run throughout the nineteenth century. In 1896 the *British Medical Journal* summed up the situation that arose when there was no revaccination:

> *Primary vaccination vastly reduces the mortality from small-pox, but it also shifts the incidence of this mortality from childhood to adult life ... In the last century the survivors of small-pox in childhood were permanently protected. We substitute the less permanent protection of a single vaccination in infancy, a protection which as a rule lapses more or less with age, and the result is that there is a vast saving of human life on the whole. This saving is entirely amongst the young. More adults now die of small-pox than in the early days of vaccination.*

This was certainly true in the severe epidemic in Paris in 1825, and it led to a movement for revaccination. Württemberg was the first to introduce it, in 1829, and it was made compulsory in the Prussian army in 1834, followed by the Russian army, the Danish army and other German states; compulsory revaccination in the British army did not come until 1858, although recruits had certainly been vaccinated in the Napoleonic Wars. From about 1840, all nurses and servants at the London Smallpox Hospital were revaccinated before they moved in, and none caught smallpox. But even this evidence failed to move the National Vaccine Society, which said in 1851 that revaccination was 'as incorrect in theory as it was uncalled-for in practice'. Twenty years later Marson, who was then in charge of the Smallpox Hospital, produced convincing figures:

> In thirty-five years he had never had a nurse or a servant with small-pox; he revaccinated them when they came. In 1871, out of a total of 110 attendants at Homerton, all but two were revaccinated, and these two took small-pox. Again in 1876–7, all the attendants but one were vaccinated, and this one died of small-pox.[15]

In 1898 the British government recommended general revaccination, but Edwardes, in his 1902 history of smallpox and vaccination, argued strongly that 'recommending' was inadequate. He pleaded for 'obliga-tory revaccination in school-age ... because we *know* that revaccination is necessary'.[16]

It was easy to insist on vaccination and revaccination for armies, which were by their nature under orders from the state, but the ques-tion of legislation on vaccination for civilians emerged quite early and was more complicated. John Haygarth, in his 1793 pamphlet on exter-minating smallpox through inoculation, had considered that a policy of compulsion and fines would be necessary, even if it meant 'inva-sion of personal liberty'. The first ruler to attempt to make vaccination compulsory was Napoleon's sister Marianne Elisa of Lucca, in 1805, though she did not solve the problem of how to enforce it. In 1806 de Carro used the *Bibliothèque britannique* to plead the cause of compulsion, and wrote that the Slavs enforced it by sending the surgeons around with a pair of Hungarian soldiers.[17] Well-meaning legislation started appearing in various countries of Europe – Bavaria in 1807, Denmark 1810, Norway 1811, Russia 1812, Sweden 1816, France not until 1902

– but compulsion needed an infrastructure that often did not exist. Russia in particular was understandably preoccupied with repelling Napoleon, and was not able to vaccinate all the babies who registered.

A related question, once vaccination was well established, was the need to ban variolation. As early as 1802 de Carro, pleased that it had become hard to find inoculators in Austria, had added that inoculation should be shunned, like goods coming from a plague-ridden country: 'it is inconsistent for a government to encourage vaccination and not forbid inoculation'.[18] Marianne Elisa had understood the problem, and had included the ban in her legislation, but in Britain it did not happen without a struggle. In 1806 Jenner wrote to Dunning: 'What havoc the anti-vaccinists have made in town by the re-introduction of variolous inoculation! It is computed that, since April last, not less than 6,000 persons in the metropolis, and the villages immediately in contact, have fallen victims to smallpox.'[19] The following year, in the debate on the second parliamentary grant to Jenner, the question of banning inoculation was graphically raised by one Member: 'I think that this legislature would be as much justified in taking a measure to prevent this evil by restraint, as a man would be in snatching a firebrand out of the hands of a maniac just as he was going to set fire to a city.' [20] It did no good. 'You will be sorry to hear the result of my interview with the Minister, Mr Percival', Jenner wrote despairingly to Lettsom. 'Alas! all I said availed nothing; and the speckled monster is still to have the liberty that the Small Pox Hospital, the delusions of Moseley, and the caprices and prejudices of the misguided poor, can possibly give him.'[21] But the College of Physicians were doing their best for the cause, and reported to Parliament:

> *Till vaccination becomes general it will be impossible to prevent the constant recurrence of the natural smallpox by means of those who are inoculated, except it should appear proper to the Legislature to adopt, in its wisdom, some measure by which those who still, from terror or prejudice, prefer the smallpox to the vaccine disease, may in thus consulting the gratification of their own feelings, be prevented from doing mischief to their neighbours.*[22]

Various attempts at legislation against variolation and for quarantine and for vaccination were made in the English Parliament in 1808, 1813 and 1814, but all failed.

An epidemic period from 1837 to 1840 brought smallpox deaths to London on a scale that had not been seen since vaccination began. It was scandalous, William Farr wrote in *The Lancet* in 1840, that even the new Vaccination Act was so inadequate that five children a day were dying of smallpox in London. Imagine, he said, the outrage if five children a day were thrown off London Bridge; the smallpox deaths could be stopped in a week if every unvaccinated person in London was vaccinated. The Act did, though, offer free infant vaccination – the first free medical service in Britain.[23]

In response to a report from the Provincial Medical and Surgical Association (later to become the British Medical Association), the 1840 Act also banned variolation. All the medical profession agreed that this was an entirely good move – an Irish doctor wrote that he had 'never known variola [smallpox] to make its appearance, excepting after the visit of the inoculator'.[24] But the provision for vaccination was much criticised. Responsibility for vaccination was given to the Poor Law Commissioners, who were to employ vaccinators at one shilling and sixpence for each operation (to be paid out of the rates). It would be hard to think of a more unpopular system. The poor who were not vaccinated – whether simply from poverty, or from ignorance, prejudice or apathy – hated having anything to do with the Poor Law Commissioners, with their overtones of Dickensian workhouses. The pages of *The Lancet* that year are full of letters from doctors who did not want to work through such contracts and were concerned about the system of appointment of vaccinators. One shilling and sixpence was condemned as insulting pay which would attract only hopeless vaccinators (who, unlike vaccinators on the Continent, did not have to be qualified); some doctors said they would rather vaccinate the poor without a fee than accept such an amount, and that the fee for those who could afford it should be five shillings. There had been talk of compulsion but, as the editors wrote, 'even a wholesome tyranny is odious' and persuasion and encouragement would, they thought, be effective in time. Making vaccination free for the patient had taken away one barrier to compulsion, but there were plenty of problems left. It was not just the question of interference with individual liberty – the liberty of exposing a child to smallpox – there was also a danger that associating the idea of vaccination with fines and punishment might do more harm than good.

It was soon obvious that the provisions of the 1840 Act were inadequate. Two-thirds of the smallpox deaths in the following decade were still children under five, [25] – many children were being vaccinated far too late if at all. Urged by the Epidemiological Society, who argued that the greatest risk came from public apathy, the government introduced a further Bill in 1853, proposing that all infants should be vaccinated before they were four months old, and they committed themselves to the idea of fines or even imprisonment for parents who refused. Again the government seemed to be trying to do the right thing, but in the wrong way. Instead of paying doctors and giving them the responsibility to carry out the scheme, the Bill was still leaving all organisation to the Poor Law officials, and the threats made the whole scheme unpopular. As part of the discussion over the Bill, the Epidemiological Society produced a report on the way other countries were tackling the problem. South Africa, and many European countries, even if they did not have penalties for failing to vaccinate, made it in effect compulsory by demanding a certificate of vaccination for schooling, or poor relief, or apprenticeship or even (in Bavaria, Prussia and Denmark) for marriage. The result was, the report said, 'that the proportionate mortality from small-pox in England and Wales is considerably more than double what it is in any country in which vaccination is compulsory'. And the European figures would be even better, it added, if public vaccinations were available several times a year, and certificates were needed at an earlier age, so that infants could be better protected.[26]

The doctors protested about the terms of the 1853 Bill as they had in 1840 – the role of the Poor Law Commissioners, the mean payments, and (in a way that sounds only too familiar) that no extra pay was being offered for all the extra paperwork involved – but when the Bill passed into law they did the best they could. The annual average death toll from smallpox in England and Wales, nearly 12,000 in the three years between 1838 and 1840, halved after the 1840 Act and halved again after 1853. John Simon, who in 1858 became England's first Chief Medical Officer, suggested that the next step should be to make compulsion more effective, not by unworkable fines and threats but by visits from local officers, and by ensuring that vaccination was competently done. Incompetent vaccinators not only failed to give proper protection, but sometimes operated dangerously on children who were not in a fit state to take it. In his Petition to the House of Commons (1856), Marson (of the Smallpox Hospital) wrote that

> It should be remembered that no authorised system of vaccination has been established in England. All persons – medical men, clergymen, amateurs, druggists, old women, midwives &c – are allowed to vaccinate in any way he or she may think proper, and the persons operated on are considered to have been vaccinated.[27]

Vaccination, he said, was organised far better in Sweden and Denmark than in England.

The neglect of vaccination in Scotland, and the smallpox among unvaccinated children there, led the Professor of Medicine in Edinburgh to comment in 1857 on the contrast with 'the savages of new Zealand', who 'practice vaccination willingly and generally, and in the opinion of the English medical officers in that part of the world, keep their country absolutely free from that disease'.[28] He mentioned, too, how successful vaccination was being in various parts of Europe, and in India, Ceylon and Mauritius. Jenner had written to James Moore in 1812 on the way smallpox was being conquered in Mauritius and at the Cape of Good Hope. 'But why', he had also written despairingly, 'should any one perish by this disease here, when there are so many examples before us of its being rooted out in every town, city, and district where vaccination is practised universally?'[29] Some of these reports turned out to have been over-optimistic.

The defeat of Napoleon had checked the progress of vaccination in France. French doctors were frustrated at the way the government was now refusing to get involved, and at the complete indifference of the poor. The worst affected areas were in the big cities in the north, particularly round Paris and Lille.[30] But French nuns, back after revolutionary times, vaccinated the children and the sick in their charge in France and in Italy, and by offering vaccination to the poor they helped to overcome the prejudice associated with fear of doctors. With an echo of the early inoculations by Lady Mary's 'set of old women' in Turkey, a school of 'wise-women' (*hakimas*) was set up in Egypt in 1832. After four years' apprenticeship in their various duties, they travelled around on donkeys as midwives and vaccinators.

In his *Papers relating to the History and Practice of Vaccination*, produced in 1857, a year after the end of the Crimean War, John Simon does not mention the war at all. Notoriously, there were far more deaths in the Crimea from illness than from battle, but the villains were cholera, scurvy and typhus – smallpox does not seem to have played any significant part.

———

In North America in the first half of the nineteenth century, vaccination was reducing smallpox among the white population, particularly in the northern states, though there were periodic outbreaks both in the cities and rural areas. Figures from the Civil War armies show how different the situation was in the black population; 10 per cent of the black troops but less than 3 per cent of the white troops in the Union army caught smallpox,[31] and the disease was far more widespread in the south.

Massachusetts had a long tradition of isolation and quarantine, and an effective law enforcing it, so there were no deaths from smallpox there from 1816 to 1824, and only thirty in the following twelve years. Benjamin Waterhouse had written in 1801 that:

> *In New England, the most democratical region on the face of the earth, the priest, the magistrate, and the people, have voluntarily submitted to more restrictions and abridgements of liberty, to secure themselves against that terrific scourge, than any absolute monarch could have enforced.*[32]

Unfortunately, the Massachusetts law was repealed in 1838 (just as the immunity of the population was weakening), and the death toll shot up to nearly 1,500 in the following twenty-two years. The state redeemed itself slightly in 1855, by passing the first mandatory school vaccination law, though this gave no protection to children under school age, and in an epidemic in 1859 40 per cent of the 318 victims were under five years old. A letter forwarded to *The Medical Gazette* from Hartford, Connecticut, in 1839, had described attempts to introduce compulsion there:

> *In some of the large cities are vaccine institutions, where the poor are vaccinated gratuitously. In some of the states, as in this, the law requires the town authority to adopt measures for the vaccination of the inhabitants twice a year, and those who refuse to have their children vaccinated are liable to be fined; but this law, I believe, is wholly inoperative, though the town authorities occasionally hire a physician to vaccinate all who wish it.*[33]

This same letter gives an alarming hint of the far worse situation of the native Americans:

> *Still, we have had but very little small-pox among the civilized inhabitants, since the introduction of the vaccine disease. Cases out of the large commercial places are quite rare. During the last year it has prevailed to a frightful extent among the Indians in our western country. Several*

> *thousands perished in a short time, and no doubt many more would have done so, but for the prompt efforts of the army surgeons, who were dispatched by our government to their relief with vaccine virus.*

For many American Indians, vaccination became associated with their compulsory resettlement in the west and, as the painter George Catlin wrote, 'They see white men urging the operation so earnestly they decide it must be some new mode or trick of the pale face by which they hope to gain some new advantage over them.'[34] Smallpox was still being spread by the movement of fur traders and gold-rush pioneers, and flourished appallingly in the cramped and insanitary villages – it has even been suggested that in 1831 fur traders deliberately infected the Pawnee Indians.[35] There is a terrible list of suffering tribes, with possibly as many as 300,000 deaths among the Indians of California in the epidemics of 1836–40.[36]

The situation in Latin America was not much better, with smallpox epidemics persisting depressingly throughout the nineteenth century. It seems to have hit everywhere – in Mexico, Argentina, Bolivia, Uruguay, Chile, Brazil. Claude Lévi-Strauss wrote of a Brazilian ambassador in Paris as late as 1934 who had tried to 'divert attention from the favourite pastime of the men of his parents' generation, and even of the period of his own youth, which was to collect garments of smallpox victims from the hospitals and hang them, together with other gifts, along the paths still frequented by the Indian tribes'.[37]

At the end of the nineteenth century, though, after a series of epidemics, there was encouraging news from Puerto Rico. In 1898, soon after occupation by the United States, about 80 per cent of the island's million population were vaccinated; within a year, smallpox was eradicated.[38] By that time there were vaccine institutes in several Latin American countries, and soon most of Latin America had compulsory vaccination laws, even though the vaccines were sometimes unreliable and the laws rarely enforced.

Slight news of progress came in from the East. After its hopeful start, vaccination spread only slowly in China – partly, it has been

suggested, because the Chinese tradition of inoculation by 'introducing a dried pock scab on a lucky day into one of the nostrils'[39] did not
lead to an acceptance of vaccination as easily as the Western tradition
of inoculation in the arm, and partly because of the opposition of the
traditional inoculators themselves. Medical missionaries tried to establish centres in both Hong Kong and Shanghai in 1844,[40] though they
could not help more than a small part of this vast country.

In India, in spite of the early success of Elgin's efforts, and the
grateful subscriptions that had been sent to Jenner, vaccination was still
practised on too small a scale to prevent severe epidemics until much
later in the century. Vaccination had been introduced to Siam (now
Thailand) in 1840, with vaccine scabs sent out from Boston – scabs,
though less effective than other forms of dried vaccine, survived the
heat of the journey better. Throughout the century there were terrible
epidemics and very little vaccination in all of Indochina.

So there was still plenty of smallpox, but it was clear that the way
forward was more vaccination and revaccination, and the courage to
tackle the problems and politics of compulsion. John Simon's 1857
report concluded that:

> By vaccinations, properly administered, the once enormous fatality of
> small-pox may be reduced almost to nothing ... The vaccinations of
> Europe are now counted annually by millions. It may be vain to hope
> that every lancet shall be used with equal skill and equal carefulness, or
> that all populations shall be equally anxious to render those operations
> successful; but medicine at least has contributed her share, in showing
> that – subject to those conditions small-pox needs cause no further fear,
> nor its antidote be accepted with mistrust.[41]

11

A hundred years on

In 1896, the *British Medical Journal* celebrated the centenary year of Jenner's classic experiment with a fifty-page review looking back at the achievements of vaccination world-wide, and the Royal Commission on Vaccination published the report which led to the 1898 Vaccination Act.

There were some astonishing figures. In England, the annual death toll from smallpox between 1850 and 1869 had been nearly 4,000 (196 per million living), and in the decade 1885–94, in spite of the growth of crowded towns and an increasingly mobile population, this had fallen to an annual average of 743 (26 per million living).[1] There was a similar story wherever vaccination was enforced, and even better figures in Prussia where there was compulsory revaccination of schoolchildren – down from 272 per million living, to only 4.4. A battery of statistics comparing the state of regulations all round Europe effectively dealt with the arguments that all this could possibly be the result of 'improved sanitation' or 'natural decline of smallpox'. There was a close relation between the amount of vaccination and the decline in the smallpox death rate in every country – down to around a tenth of the pre-vaccination rate when vaccination was introduced, down again by around a third when compulsion came in, better still when it was enforced, and approaching zero when revaccination was added.[2] As John McVail, President of the Sanitary Association of Scotland, wrote: 'vaccination has been the direct means of saving more lives and preventing more misery than any single discovery that has ever been made in the history of humanity'.[3]

Japan was a newcomer in this survey of vaccination. Vaccination had come late – to Nagasaki in 1849, by the invitation of a Japanese nobleman who had learnt through the Chinese of Jenner's pamphlet and had asked the Dutch for 'seedlings'.[4] Besides its obvious importance in fighting smallpox, vaccination played a big part in pioneering the introduction of Western medicine. It was included in the central service for European medicine set up in Tokyo in 1858, a service that became a centre of Western medical research and the foundation of the Faculty of Medicine of the University of Tokyo.[5] Vaccination was adopted with thoroughness, becoming compulsory, together with revaccination, in 1876. A year earlier the Mikado had been vaccinated, during an epidemic which killed the Emperor of China.

It took a long time for the practice of vaccination to penetrate inland in China, though it was badly needed. Professor Parker, writing in the *British Medical Journal* in 1907, said that even when he was in Peking thirty-five years earlier, 'pock-marked people were the rule rather than the exception' and that parents were unwilling for their child to marry anyone who had not yet had smallpox; south of the Yangtse, he added, pockmarks were rare.[6] Dr Morrison, a traveller in western China at the end of the nineteenth century, reported that traditional inoculation was still being practised, and that 'Infanticide is hardly known in that section of Yunnan of which Tali may be considered the capital. Small-pox kills the children.'[7] Variolation in fact was still used in China almost as much as vaccination in the early part of the twentieth century, and occasionally, in rural areas where vaccine was in short supply, right up to 1965.[8] This was true too in many parts of Africa, Afghanistan and parts of Pakistan. Variolation had been made illegal in British India in 1870, but it continued to be practised from time to time in the princely states.

Travellers commented on the pockmarked population of Korea. In Tibet, the belief that smallpox victims were sent to hell added to the horror;[9] Lhasa, 'after a visitation of the disease, remained unpeopled, a city of the dead, for three years'.[10] Two mid-nineteenth-century explorers planned to introduce vaccination to Tibet, hoping that in doing so they would not only help medically, but also encourage conversion to Christianity; but they gave up the attempt, and vaccination seems to have reached Tibet only in the early 1940s.[11]

Robert Pringle, superintendent of vaccination in the North-west Provinces of India, wrote in 1869: 'If cholera carries off hundreds every year, if the victims of famine were to be counted by thousands, these are but infinitesimal quantities beside the frightful devastation caused in India by small-pox.' And famine could spread smallpox by causing movements of populations in search of food, while smallpox in turn could cause famine by attacking the farmers.[12] Pilgrims wandering from shrine to shrine brought smallpox to villages, where as many as 80 per cent of the very young patients would die. In Bengal, Pringle said, a system of native superintendents each looking after ten or twelve vaccinators was bringing some improvement.[13] This was noticeable throughout most of India by the last three decades of the nineteenth century, when *The Imperial Gazeteer of India* was able to report that deaths from smallpox had halved. The figures might have been better still if vaccination had been more thorough. 'The Vaccination Act of 1880, as amended in 1909, is intended to give power to enforce compulsory vaccination in certain areas only,' India's Public Health commissioner said. 'It has not been considered practicable to enforce this in India generally.'[14]

The British force which set out under Sir Garnet Wolseley to fight in Ashanti, Ghana, in 1873 were vaccinated on board ship, and when they got there found little opposition from Africans who were severely weakened by smallpox and fever.

Smallpox was never endemic in Australia, with its scattered population, but it played an important part in nineteenth-century Australian history. 'Between 1780 and 1870 smallpox itself was the single major cause of Aboriginal deaths', Judy Campbell has written in her detailed study of a century of diseases in Aboriginal Australia.[15] Like the native Americans, the Aborigines were unprotected, living in small

22. Vaccination on board ship, for the Ashanti expedition, 1873

close-packed groups with 'chains of connection' and the habit of visiting neighbouring tribes, and like them they thought that smallpox was
brought by evil spirits, so they could only abandon the sick and flee in
terror.

The white population of Australia was largely immune to smallpox, either from vaccination or from previous infection, and they did
start efforts to vaccinate Aborigines. We hear first of a few vaccinations done by a settler in Bathurst, New South Wales, in the 1820s,
and then of more by army and colonial surgeons there; in 1839 the
Governor ordered that vaccination should be provided for the 'Native
Inhabitants' in Adelaide; in the 1850s there were plans for vaccinating,
and for getting good stores of vaccine. The heat made it difficult to
preserve vaccine supplies – and it would probably have destroyed the
bottle of 'variolous matter' carried by the surgeons of the 'First Fleet'*
(for emergency inoculation should smallpox break out on the journey),
matter which has sometimes been accused of causing the Australian
epidemic of 1789.[16] British settlers brought tuberculosis and measles to

---

*The fleet of eleven ships bringing convicts to Botany Bay in 1788.

Australia but, as we have seen, they probably were not guilty of being the first to bring smallpox – and they certainly brought vaccination.

The American Civil War led to a smallpox crisis. Following successful experiments in Europe, vaccine had recently been produced by inoculating cowpox into a herd of cows near Boston, but when an epidemic broke out in the South, supplies of reliable vaccine ran out, and new stocks from England were held up by the blockade. In the general panic, rather than wait for health officials, the soldiers themselves started vaccinating arm-to-arm, believing that the bigger the wound the greater the protection. There were some terrible results, made worse by infections of syphilis and tetanus, while troops, prisoners of war, and freed slaves carried smallpox on to the North. In a notorious camp at Andersonville in Georgia, where hundreds were suffering from scurvy and gangrene, vaccinations were catastrophic. Rumours spread that the surgeons were deliberately poisoning their patients by vaccinating them, and when the war ended the captain in charge of the camp was hanged.[17] At one time in Georgia there was such shortage of vaccine that the doctors went back to variolation.

Among the many Americans who, it seems, had never been vaccinated was Abraham Lincoln.[18] In the evening of 20 November 1863, after delivering the Gettysburg Address, he had a severe headache and fever, the start of a relatively mild attack of smallpox which kept him out of action for nearly four weeks; his valet caught the infection and died.

The victory of the North brought the end of the importation of slaves and of the smallpox that sometimes arrived with them. But new problems came with the increase of unvaccinated immigrants from various parts of Europe, and with the growing network of railways carrying infections around the country. An effort was made to revive old regulations: in the 1880s, in Massachusetts and in Atlanta all schoolchildren were vaccinated; all railway workers in Texas were advised to be revaccinated; in 1907 vaccination was made compulsory in schools for Indian children. By the end of the nineteenth century the epidemics which had been breaking out all across the country were under control.

The United States was ready to join in the Jenner centenary celebrations.

There were Jenner celebrations in France and in Germany too. In Russia the commemoration was postponed to a date in October, the anniversary of Dimsdale's variolation of Catherine the Great. This in no way implied disapproval of vaccination. Russia had banned variolation as early as 1805 – the first major country to do so – and young Russians were being encouraged to become trained vaccinators by promises of exemption from military service and from taxes. All Jenner's works on vaccination were produced in Russian in the centenary year.

There was shamefully little celebration in England – the only reference in the 1896 *Times* was to the Russian commemoration. And the statue of Jenner that had been put on a plinth in Trafalgar Square in 1858 had been demoted four years later to Kensington Gardens to make way for military heroes.

> I saved you many million spots,
> And now you grudge one spot for me

was *Punch's* comment. Not everyone grudges it still. The recent debate about what should occupy the one vacant Trafalgar Square plinth has moved writers in *The Lancet* to plead that Jenner be reinstated.[19]

If the central story of the battle against smallpox in the nineteenth century was one of progress, it was not steady progress. There were still epidemic years, and in 1870, the start of the Franco–Prussian War, all Europe had a shock.

The war had come at a critical moment for France, with an outbreak of smallpox just starting. A third of the population was not vaccinated, troops were moving around the country, thousands were fleeing Paris – there was bound to be disaster. It has been estimated that in 1870–71 smallpox killed between 60,000 and 90,000 Frenchmen including more than 23,000 soldiers; and French prisoners of war

23. *Russian lithograph, from the 1920s, contrasting pockmarked and blinded peasants with the new vaccinated generation*

carried smallpox to Germany. The German army of over 1 million, well vaccinated and regularly revaccinated, lost only about 450 men to smallpox, but the only compulsory vaccination for German civilians at that time was for infants, and this was only in the southern states. The German Empire, almost free from smallpox, was vulnerable, and at least 162,000 civilians died. French refugees spread the epidemic to the Netherlands, Britain and Sweden; French soldiers spread it to Switzerland; 141,000 died in Austria; European immigrants took it to the United States. Worried that it would reach South Africa, the Cape government advised the extension of vaccination. In England the death toll from smallpox shot up to 23,000 in 1871 and 19,000 in 1872. In all, probably half a million Europeans died.[20] These figures are terrible, but Edwardes has compared them to pre-vaccination figures. If, he argued, an epidemic year is one in which a tenth of the deaths are from smallpox, then in London in the forty-eight recorded years of the seventeenth century there were ten epidemics (possibly twenty in the whole century), in the eighteenth century there were thirty-two, and in the nineteenth century not even 1871 would qualify.[21]

Where there had been little vaccination, the smallpox deaths in 1871 were mostly among children; where only infant vaccination had been compulsory, the deaths were mostly among adults. In Bavaria it was particularly noticeable that nearly all the smallpox cases (around 30,000, or one person in every 150 of the population) were in an adult population who had been vaccinated only in infancy, leading to 800 adult deaths and highlighting the need for revaccination. Seeing how terribly vulnerable their populations still were, the British Parliament strengthened the Vaccination Acts by appointing local inspectors, and the Germans ruled (more effectively) that all infants must be vaccinated before they were two and revaccinated about ten years later.

The strengthening of the Vaccination Acts roused a growing rebellion in England. Most doctors realised that the weakening of immunity was a reason for revaccination, but rebels saw it as a further argument against vaccination and a reason to abandon it altogether. In his belief that smallpox could soon be conquered, Jenner had laid himself open to accusations of excessive optimism; when he had correctly deduced the reasons for occasional failures, sceptics had tended to mock the excuses; as it became clearer that he had been wrong in claiming that vaccination gave permanent protection, sceptics had more ammunition. It has been suggested in Jenner's defence that he could easily have misinterpreted the apparent permanent immunity of the milkmaids – he may not have realised that they were in effect recently vaccinated adults, and so were in a more protected state than adults who had been vaccinated only in infancy. Jenner was not alone in his optimism; Luigi Sacco never doubted the permanence of vaccination.

John Simon, in his 1857 report, tried to explain occasional smallpox in vaccinated subjects by suggesting that 'certain original properties of the vaccine contagion have very generally declined, after its long successive descent from the cow'. He added:

> *Frequent failures in vaccinating not only disappoint and annoy both parties concerned, not only discredit the operation and the operator, but likewise too often lead to an ulterior evil. Ignorant persons look rather to the mere doing of vaccination than to its success; and it constantly happens that children who have been thus nominally vaccinated are in fact left with no further attempt to secure them against smallpox.*[22]

In fact the failures must have been due to poor technique in vaccination or in storage of the vaccine rather than any slow progressive deterioration. But this vague theory set off an exotic series of experiments to reactivate the vaccine or to find new sources of animal vaccine.

A much milder form of smallpox started appearing towards the end of the nineteenth century. Caused by *variola minor*, it was fatal in less than 1 per cent of cases, left the patient immune to either form, led to far less scarring, and did not attack those who had been vaccinated. It arrived in Pensacola, Florida, in 1896, infecting more than fifty in the city and many in the surrounding country, but none died. It may have come from the Caribbean, where it had been described about thirty years earlier, or from South Africa, where it had been described the previous year. Something like it may possibly have been around longer – Jenner himself wrote in the 'Inquiry', of 'a species of Small-pox' that had appeared in Gloucestershire 'of so mild a nature, that a fatal instance was scarcely ever heard of', and Joseph Needham even suggests that the weak form of smallpox selected for inoculation in seventh-century China may have been caused by variola minor.[23] Since patients infected with variola minor did not feel very ill, they were often up and about, and at the end of the nineteenth century the infection soon spread around the United States and Canada – there are cases of touring actors cheerfully ignoring it and leaving a trail of smallpox wherever they went.[24] So although the case fatality was low, the number of smallpox cases rose. The first recorded instance in Britain was traced to a contaminated parcel sent from Salt Lake City to a Mormon conference in Nottingham in 1901, and in the same year there was a case in Stockport caused by infection in cotton arriving from Texas; it does not seem to have become endemic in Britain until 1919, though John McVail, in his Milroy lectures to the Royal College of Physicians, considered that both types of smallpox were occurring simultaneously in the first two decades of the century.[25] In Latin America, after the relentless recurrent epidemics of variola major throughout the nineteenth century, the milder disease started to take over, though there were still severe outbreaks of variola major in Peru and Mexico in the 1940s.

—

Vaccination of course did have its dangers. Apart from the small risk of transferring other infections, there was a risk that the virus in the vaccine could itself be transferred by contact with any scratch or burn or other open skin lesion, though not by inhalation. In children with eczema, vaccination – or even accidental infection from other vaccinated children in a children's ward – could lead to a generalised eruption of a 'malignant and fatal kind'.[26] Very rarely, too, vaccination could lead to a life-threatening progressive spread of the virus or to a dangerous encephalitis.

Elizabeth Garrett Anderson, reviewing the Royal Commission's

1896 report, quoted a possible figure of fifty deaths a year from vaccination in England and Wales (one in 14,000 vaccinations).[27] The
Royal Commission had been dismissive, saying such figures while 'not
inconsiderable in gross amount, yet when considered in relation to the
extent of vaccination work done, they are insignificant'. The figures
were disturbing though, and in a climate of worry and doubt it was
easy for the anti-vaccinists to build up a following – think how, in spite
of convincing scientific evidence for its safety, public feeling gets stirred
up today about the measles, mumps and rubella triple vaccine (MMR).
The nineteenth-century public had far less information and the authorities often acted with little sympathy – there are even stories that parents
whose children died after variolation were convicted of manslaughter.[28]
'It is a sentimental fad', the *British Medical Journal* wrote firmly, 'to talk
about personal liberty where the health and lives of our neighbours are
concerned in our condition as to susceptibility to small-pox.' Coercive
legislation and repeated punishment only strengthened the opposition
and brought sympathy to the persecuted. In 1880 Walter Hadwen, a
young doctor in Somerset who was to become prominent in the anti-
vaccination movement, refused to have his own baby vaccinated; four
prosecutions cost him £50 in fines.[29] In 1897 Parliament was told of
a protester who had been prosecuted sixty-three times, had paid more
than £42 in fines, and had then been sent to prison in default of further
fines.[30]

The English Anti-Vaccination League had been founded after the
first attempt to make vaccination compulsory by the Act of 1853; the
French had followed suit in 1866, and the Americans in 1879 – protests in America were particularly strong among the immigrants. The
movement reached South Africa too, with a pamphlet in Durban even
claiming that the vaccinated suffered more severely from smallpox than
the unvaccinated.[31] In Quebec, worries about infection with syphilis
had caused resistance to vaccination in the 1870s, particularly among
the French-speaking population, and there were riots when Montreal
authorities tried to control an epidemic in 1885 by enforced vaccination
and isolation; the Health Offices, the Chief of Police and the office of

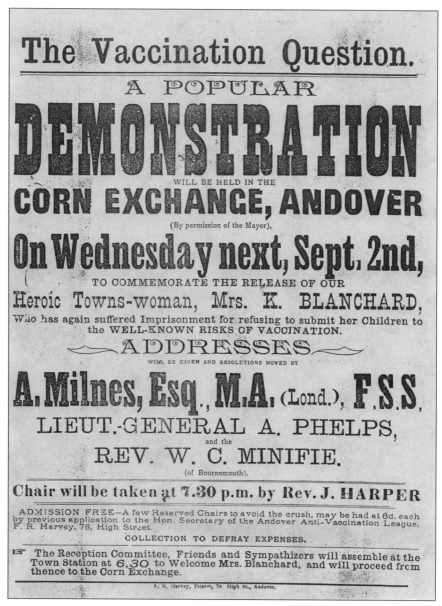

24. Anti-vaccination campaign poster, 1896

the *Montreal Herald* were all attacked, and troops were called out. As a result of the anti-vaccination prejudice many were unprotected, and over 3,000 died in that last major Canadian epidemic. *Harper's Weekly*

25. Resistance to vaccination in Montreal, 1885

described and illustrated the pathetic siege of one household, the men forcibly restrained as the children were bundled off to hospital. Imprisonment, fines and heavy-handed enforcement did not help the cause of vaccination.

Some joined the anti-vaccination movement because of an understandable dislike of compulsion, but it also collected assorted followers who objected to vaccination under any circumstances. Elizabeth Garrett Anderson, analysing this assortment, pointed out that none of them had ever had practical contact with smallpox, that there was a 'large and impenetrable body of cranks', and that there was also a 'great body of uneducated people who are against vaccination because it gives them a little trouble, some restless nights, and because they think the risk of smallpox is a remote one'. She had some sympathy for this third group, although they were 'only too ready to welcome the mischievous doctrines of the anti-vaccinist orators'.[32]

As a doctor, Walter Hadwen did not lack practical contact with smallpox, but he opposed what he called 'Jenner's cult', and in support quoted the great reformer Edwin Chadwick, who had rashly lumped smallpox in with other diseases that could be tackled by improved sanitary conditions:

> *Smallpox, typhus, and other fevers occur on common conditions of foul air, stagnant putrefaction, bad house drainage, sewers of deposit, excrement sodden sites, filthy street surfaces, impure water, and overcrowding.*[33]

No doubt overcrowding would help many infections to spread, but Hadwen and his followers failed to see that more than sanitary reform was needed in the battle against smallpox.

In spite of the overwhelming evidence of the effect of vaccination on the death rate from smallpox, there were other prominent dissenting figures. The writings of two of them in particular seemed to give the anti-vaccination movement some credibility at the time: Charles Creighton, in his *History of Epidemics in Britain*, published in 1894, contrived to argue that vaccination was 'irrelevant' to the history of smallpox, and warned of the danger of transmitting syphilis. He had written a hostile article on vaccination for the 1888 edition of the *Encyclopedia Britannica*, and thought Jenner a charlatan;[34] he 'declined to discuss modern statistics' in his evidence to the Royal Commission on Vaccination, 'having already made up his mind'.[35] Edgar Crookshank, professor at King's College London, wrote in his *History and Pathology of Vaccination* (1889) that smallpox should be attacked by variolation in the way that had been proposed by Haygarth in 1793 in his 'Sketch of a plan to exterminate the casual smallpox from Great Britain'. According to Crookshank, 'Inoculation of Cow Pox does not have the least effect in affording immunity from the analogous disease in man, syphilis, and neither do Cow Pox, Horse Pox, Sheep Pox, Cattle Plague, or any other radically dissimilar disease, exercise any specific protective power against Human Small Pox.'[36]

No one of course was claiming that vaccination gave immunity from syphilis, and much of the rest of his statement is simply untrue, but the anti-vaccinists voiced a genuine worry over infections that might be passed on in arm-to-arm vaccination. In the days before antiseptics, there were cases of erysipelas (a fierce red streptococcal skin infection) traced to 'reprehensibly careless' vaccinating. Killick Millard, the Medical Officer for Health for Leicester at the beginning of the twentieth century, cites an outbreak in 1876 when, out of a series of sixteen children vaccinated, eleven developed erysipelas and six of them died.[37] Cases of syphilis had been seen following infected vaccinations in the American Civil War and in Quebec. Congenital syphilis is not always apparent in very young infants, so there was the extra hazard that vaccine material could unwittingly be taken from the arm of an infected infant – this happened on about fifty occasions in Europe between 1800 and 1880 (out of 100 million vaccinations), and it led to an estimated 750 syphilis infections.[38] As Jurin had pointed out 150 years earlier, 'People do not easily come into a Practice in which they apprehend any Hazard, unless they are frightened into it by a greater danger.' To many, the death rate from smallpox was now so low that it did not seem to be the greater danger.

Some of the opposition was less rational, and some pamphlets were not much less hysterical than in the days of Rowley's 'ox-faced boys'. Here is *The Vaccination Inquirer* for March 1914:

> *Many of us would prefer an attack of smallpox to vaccination, all of us would prefer the risk of smallpox to the certainty of vaccination. Smallpox is a natural disease running a known course ... Vaccination is a loathsome disease of uncertain origin, artificially transmitted, through various beasts and capable of setting up a variety of repulsive, dangerous, and even fatal affections.*[39]

The anti-vaccinators appealed to the public to 'Think of the unparalleled absurdity of deliberately infecting the organism of a healthy person in this day of sanitary science and aseptic surgery with the poisonous matter obtained from a sore on a diseased calf.'[40] The idea of such 'absurdity' being imposed by law horrified many. It was, Bernard Shaw said, 'a particularly filthy piece of witchcraft'.

A surprising figure in this scene is Alfred Russel Wallace, who frayed his evolutionary laurels by an extraordinary attack on vaccination and on all the 'useless and dangerous' statistics that had been produced in its support. 'Vaccination is quite powerless either to prevent or to mitigate small-pox,' he pronounced. He then went on, 'with an amount of inaccuracy which is at once amazing and lamentable'[41] to call it 'an operation which has admittedly caused many deaths, which is probably the cause of greater mortality than smallpox itself, but which cannot be proved to have ever saved a single life'.[42] (Charles Darwin, on the other hand, duly entered his children's vaccination dates in the family bible.)

—·—

The town of Leicester played a prominent part in the anti-vaccination movement and the resistance to compulsion. In 1885, out of about 100,000 inhabitants, 4,000 were awaiting prosecution for failing to vaccinate their children, and 20,000 joined a protest meeting, burning a copy of the Vaccination Acts. Since the law was administered by elected Poor Law Guardians, the way to frustrate the law was to elect Guardians opposed to it. When this was done in Leicester there were no more prosecutions, and the child vaccination rate, which had been 90 per cent in 1872, fell to 3 per cent in 1892.* Dr Priestly, appointed Medical Officer of Health for the town in 1892, believed in vaccination, but the Sanitary Committee did not allow it. Luckily there were no further major epidemics like that of 1871–2 to challenge Leicester's unvaccinated state, but in a minor epidemic over a hundred unvaccinated children caught the disease, of whom fifteen died; and there were about fifty cases, with four deaths, in unvaccinated adults. In a comparatively compact city (and surrounded by a vaccinated popula-

*The ploy did not always work. When it had been attempted in 1876 in Keighley, the seven Guardians had themselves been imprisoned in York Castle for refusing to enforce the Act.

tion), the authorities struggled to control the outbreak by notification (compulsory in Leicester in 1879, though nationally not until 1899), by quick isolation of suspected cases and by disinfection of houses. In theory, vaccination was being totally rejected, but in practice, *The Lancet* reported, all but six of the twenty-eight nurses in the local smallpox hospital were revaccinated – and of those six, four caught smallpox, one fatally.

When Killick Millard became Leicester's Medical Officer of Health he brought in a more pragmatic approach. As the Royal Commission on Vaccination had said in its 1896 report, what isolation 'can accomplish as an auxiliary to vaccination is one thing; whether it can be relied on in its stead is quite another thing'. And isolation, as the ever-practical Elizabeth Garrett Anderson pointed out, as well as being useless on its own, is far more expensive than vaccination. Millard did not reject vaccination, but thought it should be used in the context of good sanitation; and sanitation, as he saw it, was not just a question of cleanliness and a good water supply, but included the more relevant problems of preventing overcrowding, and providing enough hospital beds and proper facilities for isolation. Millard accepted that vaccination in infancy followed by revaccination, if thoroughly enforced as in Germany, would be thoroughly effective; but he did not believe such regulation to be possible in England except for groups in extreme risk such as the smallpox nurses. He doubted the value of universal infant vaccination when there was no routine revaccination, because when the immunity weakened, vaccinated subjects might suffer an undiagnosed attack of smallpox and thus infect others. While he accepted that infant vaccination would be helpful for the individual, he thought it put the community at risk – though as John McVail pointed out in his 1919 lectures, it was hard that children should be 'allowed to remain unvaccinated in order that if attacked they might have an illness sufficiently severe to make diagnosis easy'.[43] Millard relied on preventing the disease from spreading by vaccinating contacts as soon as smallpox was reported. So, given the enormous reduction in smallpox that had taken place in the previous century, he saw the practicality of something more like the 'surveillance-containment strategy' that was to emerge in the end phase of the eradication programme later in the century. In his hands the 'Leicester Method' meant 'notification, hospital isolation, disinfection, and vaccination and surveillance of contacts'.[44]

Relying on this method was less risky than it might seem. Smallpox has an incubation period of around twelve days, but vaccination becomes effective after only three to five days — so there is time for some last-minute preventive action. According to the virologist Frank Fenner, vaccination, 'if carried out during the first week of the incubation period of smallpox, might ameliorate or sometimes abort the disease'.[45] (Erasmus Darwin had even had the doubtful idea that this was true of variolation — 'Hence if the inoculation be performed 2 or 3 days after the natural infection, the patient may still take the artificial disease and escape the natural.'[46])

James Simpson, famous for introducing the use of chloroform for anaesthesia, had suggested a policy rather like the one that evolved in Leicester, as early as 1868.[47] Fully accepting the 'beneficial influence of Dr Jenner's immortal discovery' in dramatically reducing the death toll from smallpox, Simpson thought universal vaccination unnecessary, but felt that it should be possible to finish the job and stamp out smallpox from Britain altogether with selective treatment, as rinderpest (cattle plague) had recently been stamped out. Success with rinderpest had been achieved by killing all diseased cattle and their contacts — the policy the British used in the outbreak of foot and mouth disease in 2000. For smallpox, the analogous policy would be rigorous isolation and quarantine of contacts; Simpson also emphasised the importance of notification, and the need for nurses to be vaccinated. Millard put all this into practice in Leicester, and it worked. It worked too in the epidemic of 1892 in Sydney, a town where quarantine was easy, in a country where smallpox was not endemic.

The tale of two other English cities shows the impact of smallpox in different circumstances. In Sheffield, a town of around 300,000, 95 per cent of the population were vaccinated, so when an epidemic began in 1887 it was not as catastrophic as it might have been. But the arrangements for disinfection and isolation were 'rudimentary', and the one smallpox hospital, situated in the middle of the town, became 'a small-pox distributor'. The unvaccinated 5 per cent, 'an experimental population belonging to the eighteenth century, living in Sheffield under nineteenth-century conditions',[48] suffered 274 deaths (a hundred of them being children under ten), while there were only 200 deaths (including six children) in all the rest. These figures tallied with the relation between vaccination and the decline in smallpox death rate that had been reported all around Europe. There could not have been a clearer demonstration of the value of vaccination; and the fact that 200 vaccinated patients, but no one who had been revaccinated, had died reinforced the argument for revaccination.

While Leicester lacked vaccination and Sheffield lacked arrangements for isolation and quarantine, Gloucester – a far smaller city only a few miles from Jenner's home at Berkeley – lacked both. After twenty years free of smallpox, there was no provision for an emergency. The anti-vaccinists had been at work, supported by the local press, and in a population of 40,000 there were, in 1895, about 10,000 unvaccinated children. The result, in an outbreak of severe smallpox, was 2,000 cases and 434 deaths (including 42 per cent of the unvaccinated cases); 281 children died. The history of the beginning of the outbreaks here and at Sheffield supports Millard's (and Elizabeth Garrett Anderson's) view on the hazard from undiagnosed smallpox in once-vaccinated subjects. The infection soon spread disastrously to the unvaccinated schoolchildren. Hospital accommodation was hopelessly inadequate, most cases were treated at home and all attempts at quarantine broke down. Business and trade in the city were paralysed. It took compulsory primary vaccination and revaccination, carried out by six vaccinators travelling from house to house and covering the whole population, to bring the epidemic to an end.

Apart from such emergency action, revaccination was never compulsory in England, in spite of medical opinion and the undoubted success story of Prussia. Many people understood the need. Queen Victoria's daughter, the Empress Frederick, wrote anxiously to her mother*:

> ... if you could induce Moretta [her daughter] not to be so foolish about her food. Her one craze is to be thin ... If also you can persuade her to let Dr Reid vaccinate her. All the doctors say it ought to be done before going to Greece and even independently of that: from time to time one must be revaccinated.[49]

Lewis Carroll (C. L. Dodgson) shared the belief in revaccinating from time to time, being vaccinated in response to smallpox alarms in 1863 and in 1871. Wearing his mathematical hat he challenged the statistics of an anti-vaccination correspondent in a series of letters to the *Eastbourne Chronicle*.[50]

Earlier in the century, when much of the adult population had gone through either natural or inoculated smallpox, infant vaccination was all that was needed; but now those infants were outgrowing their protection. In the postal service or the police, where all employees had to be revaccinated on entry, there were no smallpox deaths, even in the epidemic of 1871. The Royal Commission concluded that revaccination 'restores the protection which lapse of time has diminished', but felt that there would be too much public opposition to any attempt to make it compulsory.

It was not only in Sheffield that smallpox hospitals, with all the coming and going that surrounded them, were noted as centres of infection. The need for isolation was tackled by a floating hospital on the Tees in 1895, by hospital ships in Bristol, and in London by the River Hospitals, set up in 1884 on a bleak and deserted stretch of the Thames at Dartford. There were three ships on the Thames, two of them able to hold 100 to 150 patients (one of these was sister ship to the *Agamemnon*, which had laid the first Atlantic cable), and the third for administration. A shore hospital

---

* Queen Victoria herself had been vaccinated as a two-month-old princess, arm-to-arm from the two-month-old grandson of Dr John Lettsom.

26. *Smallpox hospital ships on the Thames estuary, late nineteenth century*

was added in 1902, and the whole complex was served by paddle-steamer ambulances and a horse-drawn tramway. The ships were sold for scrap in 1904, but the paddle-steamers struggled on until the London County Council took over in 1930. Since the demand for smallpox nursing was variable, the hospital ships were able to send staff to help in the crisis at Gloucester – though the medical superintendent wrote scathingly that 'a municipal authority which courts an epidemic and in the consequent emergency offers a salary of £30 for competent smallpox nurses, is hardly to be pitied if it finds such nurses hard to obtain'.[51]

Faced with such opposition to compulsion as the Leicester revolt of 1885, and the practical boycott of vaccination in towns such as Gloucester, the English government took a step backwards. The Royal Commission's report led to what Edwardes, in his 1902 *Concise History of Smallpox and Vaccination in Europe*, called 'the great surrender' of the 1898 Vaccination Act. Instead of introducing revaccination, the Act allowed parents to refuse to vaccinate their children at all – to be, in the first use of the phrase, 'conscientious objectors'. The commissioners had argued that pressure and penalties only encouraged resistance, and did more harm than good. It was not clear what constituted a conscientious objector, but a quarter of a million exemption certificates were granted in the first year. By 1912, although only about half of the infants born were being vaccinated, the mild form of smallpox predominated, and the total number of smallpox deaths in England

27. Triumph of De-Jenner-ation; cartoon prophesying disaster from clause in 1898 Vaccination Act allowing conscientious objection

and Wales that year was twelve. All members of the armed forces were vaccinated or revaccinated though, and where smallpox did break out there was some revaccination; so (whether by vaccination, revaccination, or previous infection by variola minor) a considerable part of the population was in fact protected. When the National Health Service was set up in 1948, vaccination was no longer compulsory, and routine vaccination ended in 1971. Luckily as things turned out, the prophets of doom were wrong about the 1898 Act.

12

'Bring hither the fatted calf'

Right from the start there was a problem in getting supplies of cowpox vaccine. Jenner's 'Inquiry' did not appear until two years after the vaccination of James Phipps because there was no cowpox around for him to use in his search for further evidence. 'True' cowpox infections are still rare – four instances were recorded in Britain between 1960 and 1973[1] – and the virologist Derrick Baxby thinks that they probably were just as rare in the nineteenth century.[2]

'We know', Baxby has written, 'that cowpox is not found outside Europe.'[3] But has it, or something like it, never been? We have seen that in the fifteenth century the Chinese sometimes used pills made from roasted and ground water-buffalo lice in the hope of preventing smallpox. Another version of the story, 200 years later, gives a recipe for cow-fleas mixed with rice.[4] Edwardes reported that Humboldt (in Mexico) and James Bruce (in Baluchistan) had both come across the tradition that cowpox protected milkers from smallpox;[5] in 1804 the Balmis expedition reported cowpox in Mexico and Caracas;[6] in 1832 successful vaccinations seem to have been carried out from a cattle disease in Bengal;[7] in 1882 cowpox in Pennsylvania, said to be spontaneous, was used to found a vaccine farm, the forerunner of Wyeth Laboratories.[8]

Jenner's original theory was that cowpox developed when cows were milked by men who had been looking after horses and had been exposed to 'grease'. Milkmaids, he thought, had not had cowpox in the days when only women did the milking and the cows had no contact with grease; so although Jenner's discovery is often written about as if he had noticed the immunity to smallpox inoculation only in milkmaids, many of the cases cited in the 'Inquiry' are men. James Moore wrote that blacksmiths and farriers, as well as milkers, often resisted smallpox inoculation – though the doubtful immunity of three men infected by grease but not by cowpox, led Jenner to conclude that grease was more reliable when it had been through a cow.

The theory that cowpox has its origins in grease caused endless arguments. George Pearson wrote scathingly that:

> ... this conclusion has no better support than the coincidence in some instances of the prevalence of the two diseases in the same farm, and in which the same servants are employed among the horses and cows.[9]

And he agreed with Dr Parry who wrote to him that 'the assertion that the Cow-Pox proceeds from the heels of horses is gratuitous'.

It may have been the torrent of criticism that made Jenner drop this part of his theory in his later pamphlets. But the idea that horses might also be a source of vaccine did not go away. John Loy, a Yorkshire physician, reported vaccination experiments on children in 1801, some directly from a patient suffering from grease, and some from the vesicles on a cow which had been inoculated with grease; the children seem to have had rather severe reactions, but they did prove to be immune to inoculated smallpox.[10] Luigi Sacco did similar experiments, and was so convinced that the infection originated in horses that he suggested the inoculating material should be called 'equine' rather than 'vac-cine'. Jenner wrote to James Moore in 1813 that he himself had been using equine material 'arm-to-arm for these two months past, without observing the smallest deviation in the progress and appearance of the pustules from those produced by the vaccine'. Four years later he noted that he 'Took matter from Jane King (equine direct), for the National Vaccine Establishment. The pustules beautifully correct.'[11] This stock was sent around Britain to many vaccinators, including John Baron.

Crookshank described investigations into horses as a source of vaccine in France in the 1860s and 1880s.[12]

One complication is that Jenner had called the infection 'grease'; but it has been shown that in horses, as in cows, there are various similar infections, and the useful infection was not grease but another disease in horses' heels. This was first pointed out by Loy in 1801, after a series of experiments with 'matter' from horses. Jenner himself came to realise that 'the matter which flows from the fissures in the heel will do nothing. It [the effective material] is contained in vesicles on the edges and the surrounding skin.'[13] A further complication for the historian is that this useful 'horsepox' almost certainly no longer exists.*

The search for safe and available supplies of vaccine went on. New sources were found — some useful, like Sacco's Lombardy cowpox; some doubtful and experimental like the goatpox used to vaccinate an orphanage in Madrid in 1804. Jenner was sceptical when Dunning told him about the goatpox. 'Is there any quadruped that is not subject to diseased nipples?' he wrote back.[14] But whatever its source might be, the problem of preserving and distributing the vaccine was the same.

Shortly after Jenner's experiments, cowpox had been found in two dairies on the edge of London. The cowpox there could not have been derived directly from horsepox, as there had been no contact with infected horses. This was the source used by Woodville at the Smallpox Hospital, and by Pearson for the 200 impregnated cotton threads he circulated around England and sent off to the Continent. As we have seen, Jenner sent threads to his old schoolfriend John Clinch in Newfoundland, and Haygarth sent some to Waterhouse in Boston. But cotton threads impregnated with the contents of cowpox pustules and dried did not always remain effective — failures had been reported in Hanover, Genoa and Paris before de Carro had the success in Vienna that launched his career as a vaccinator.

Dried vaccine had problems, but it was easier to preserve vaccine

*In support of Jenner's original idea, though, Derrick Baxby believes that it may well be the virus most closely related to the vaccinia used today — see Chapter 13.

that way than any other. Jenner wrote that variolators had run into trouble when they had 'preserved the variolous matter ... on a piece of lint or cotton, which in its fluid state was put into a vial, corked and conveyed into a warm pocket',[15] and he realised that cowpox vaccine would have the same problem. When he went to London to arrange for the publication of the 'Inquiry', he had taken with him vaccine dried on a quill, and it was this that was used by Henry Cline for the first vaccination experiment in London. Sometimes dried vaccine was sent on an ivory point, made by comb-makers and looking like a tooth of a fairly coarse comb.

In 1803 Jenner, together with John Ring, wrote to the Lord Lieutenant of Ireland giving guidance on the 'mode of obtaining and preserving the genuine vaccine matter':

> *When fluid matter can be procured, it is always to be preferred ... But when it is impracticable to obtain fluid matter there are various modes of transmitting it in a dry state. When it is to be used within two or three days, a lancet is a proper vehicle; but if it is not used within some such period the lancet rusts ... Many safe and effectual methods of preserving Cow-Pock matter have been devised ... Some of the most common are, the taking it on thread or glass, and suffering it to dry ... Another mode ... is to take the fluid on a quill shaped like a tooth-pick ... stopped with white wax. Sealing wax is unfit for the purpose. The heat necessary for melting it may decompose the matter.*

A letter in *The Lancet* in 1840 welcomed the introduction of the penny post, saying how it would help the Smallpox Hospital to distribute vaccine.

Jenner had written that it was best to use material directly from a pustule on a human arm, but the arm-to-arm method turned out to have many problems. Apart from the major worry of syphilis, erysipelas and other infections, there were obvious difficulties in assembling teams of orphans, or in travelling around the Alps with a little child in tow to be used as a vaccine store. And to keep up the supply, vaccinated patients had to be persuaded to submit their arms to the lancet a second time; John Walker, the Resident Inoculator [vac-cinator] of the Royal Jennerian Society made matters worse by terrify-ing the parents who had brought their children back for examination: "Thou foolish woman, if thou wilt not do good to others, I will bless thy little one", and forthwith drew his lancet.'[16]

Since there were all these problems in getting supplies of vaccine from children, why not turn back to cows? There were two differ-ent approaches. One, working on Jenner's theory that cowpox was a form of smallpox, was to try to produce cowpox by inoculating cows with smallpox. Several people claimed occasional success: Gassner of Günzburg,[17] Macmichael in Egypt, Robert Ceely in Aylesbury, John Badcock in Brighton, Reiter in Munich, Basil Thiele in South Russia,[18] the Colonial Medical Committee in South Africa.[19] Ceely was lyrical in describing his results:

> by the tenth day the vesicle was commonly in its greatest beauty and
> highest brilliancy, glistening with the lustre of silver or pearl, having
> the translucency and appearance of crystal, or shining with a pale blue
> tint ...[20]

The vaccine derived from such pustules was used to immunise thou-sands, and John Simon announced proudly in his report to the British Parliament in 1857 that modern science had thus proved that Jenner was right in calling cowpox 'smallpox of the cow'.[21] But, as we shall see later, twentieth-century science has proved the interpretation of those experiments to be wrong. The explanation of their apparent suc-cess seems to be simply that the smallpox used for inoculation must, in the successful cases, have been contaminated with cowpox[22] – the reverse of the contamination occurring in Woodville's early vaccina-tions at the Smallpox Hospital.

So it is not surprising to find that a more direct approach, inoculat-ing cows with cowpox taken from a human arm, proved more fruitful. This was tried in Italy as early as 1805, and then in 1843 Negri, in

Naples, carried the process a step further by inoculating from one cow to another, scarifying the animal's side to expose a large area for the vaccine to reproduce. (In a letter to de Carro in 1799, Jenner had talked of attempts to transfer cowpox from the nipple of one cow to the nipple of another, but 'no effect has followed its application in any instance that has come to my knowledge'.[23]) Having got his infected scarified cow, Negri started the exotic practice of touring the streets, vaccinating people out of doors, or in their homes or anywhere that was convenient. The immunologist Hervé Bazin describes how a French doctor, Lanoix, travelled to Naples in 1864 to learn the technique from Negri, and returned by train with an inoculated heifer in a specially fitted guard's van padded with straw. They stopped off at Lyons to vaccinate some children and inoculate another heifer, then on to Paris where they set up successfully in the suburbs.[24] The use of calf vaccine spread to Belgium, Germany, Switzerland, Holland, Russia, and some states in the USA; and to Japan in 1874, where arm-to-arm vaccination was banned in 1891.

Picturesque scenes of crowds waiting in unlikely places to be vaccinated from tolerant calves did not take place in England. But in 1881, after visiting animal vaccination stations in Europe, the Chief Medical Officer to the Local Government Board set up a department of the National Vaccine Establishment in Lamb's Conduit Street, in London, where one calf a week was inoculated with vaccine, and 'calf-to-arm' vaccination took place, as well as the production of vaccine for distribution. This was the start of the Government Lymph Establishment, which was to become the British Institute of Preventive Medicine in 1891, and then eventually the Lister Institute.[25]

Material taken from pustules on the animal's skin – 'the vaccine pulp' – was partially clarified to produce 'vaccine lymph', still milky and contaminated with bacteria and fragments of cells. The Royal Commission of 1896 recommended that this lymph, suitably treated, should always be offered as an alternative to human lymph, since it was so much safer; with animal vaccine, there was no longer a danger of passing on other human infections, and if a heifer was young enough it

28. Paris: vaccinating in the street, 1893

was almost sure to be free from bovine tuberculosis. Occasionally there were scares with calf lymph, such as the cases of post-vaccination teta-nus in New Jersey and in Philadelphia in 1902 – which were found to be the result of secondary infections from some 'gross breach in the care of the wound', from dirty dressings or from contact with stables.[26] In England something of a two-tier system had developed towards the end of the century, the free vaccinating stations often supplying only

human lymph, while people who could afford it paid for calf. Gloucester, in the sudden crisis of the epidemic of 1895, was able to call on the
stores of the Calf Vaccine Institution of Bradford.

When lymph could be produced in large quantities it was important
to be able to store it safely, and here again Negri seems to have shown
the way. He is generally credited with being the first to use glycerine to
keep the lymph moist, the right consistency to stick to the skin, and free
from infection. As early as 1805, though, Balmis had recommended
that in the vaccination institute he set up in Manila the vaccine, from
water buffaloes, should be sent round in glycerine between glass slides
sealed with paraffin[27] – probably neither Balmis nor Negri fully understood the scientific basis for what they were doing. In 1898 *The Lancet*
reported that in 1850 R. R. Cheyne had 'mixed glycerine with vaccine
lymph … with the satisfactory result, better than any theory, of discovering that, in addition to its known property of preventing fermenta-
tion and mouldiness in vegetable substances, it had also that of keeping
vaccine lymph, an animal product, undecomposed in a fluid state for
months'.[28] By the time of that *Lancet* report Monckton Copeman,
helped by knowledge of the work of Koch and Pasteur, was able to
explain why; he demonstrated not only that glycerine kept the vaccine
in a convenient state, but also that it killed bacteria without harming
the vaccine.[29] As Elizabeth Garrett Anderson put it: 'by an intimate
admixture of lymph and pure glycerine, and by storing the mixture for
a considerable time under conditions which prevent the access of light
and air, the foreign or extraneous organisms in the lymph are gradually
destroyed and the vaccine organism only left'.[30] Glycerine also proved
useful for diluting the vaccine lymph when supplies were short – presumably because the usual dose of undiluted lymph contained more
vaccine than was necessary; the London Hospital increased the volume
available for vaccination this way during the 1870–71 epidemic.

An average calf properly prepared, Monckton Copeman claimed,
could supply material for glycerinated lymph for at least 5,000 vaccinations; before the introduction of glycerine it could have served
only for 200 or 300. Ideally, the animal used was examined by a vet

beforehand, was looked after carefully and washed before vaccination, and killed and given a post-mortem examination after the incubation period. Such treatment was expensive, and often not followed. Standards varied.

Glycerinated calf lymph became the standard form of vaccine used world-wide, and arm-to-arm vaccination in England was finally banned by the Vaccination Act of 1898. Thomas Whiteside Hime, the head of the Calf Vaccine Institution at Bradford, wrote in the Jenner Centenary Number of the *British Medical Journal* in 1896 that he had sent vaccine out successfully to Algiers, Egypt, India and tropical South America.[31] And he added that 'Public vaccinators begin to find that their stations, which had become deserted while they used humanised vaccine, soon are gladly visited by parents when they learn that calf vaccine is used.' Worries of infection were over, and it was a relief not to be pressed to offer their child's arm for the next round of patients. The same journal reported M. Hervieux, speaking at the French centenary celebration: 'the triple base of animal vaccine, revaccination, and compulsion' would, he said, ' ... defy all the assaults of smallpox, and may be proclaimed to be the grandest conquest of medical science'.

A drawback before the days of efficient cold storage was that calf lymph could be kept for only around six weeks. Improved refrigeration techniques made it possible for the Local Government Board to have half a million tubes of calf lymph in cold storage by 1912, and 'a thousand tubes could be sent to any part of the country on receipt of a telegram'.[32] Throughout the First World War, three quarters of a million doses were kept in reserve in England, and more than 7 million doses were issued.[33] At that time England, followed by other countries, started to use sheep as well as calves as vaccine providers; the Indians successfully used water buffaloes.

There were further refinements. The next move, introduced first in 1924[34] to reduce the time needed for the glycerine to do its bactericidal work, was to add a small amount of phenol (carbolic acid) to the mixture. Even so, glycerinated calf lymph was not the perfect medium. The liquid vaccine was fine in temperate climates and where there was good refrigeration, and it was successfully used to eliminate endemic smallpox from Europe and North America. But at higher tempera-tures glycerine did in fact damage the vaccine and it caused failures in tropical countries; in 1919, for example, only 7–20 per cent positive results were obtained for primary vaccination in Tanganyika (modern-day Tanzania). In Peru the problem was tackled by storing the vaccine in kerosene-powered refrigerators carried on the backs of mules.[35]

Less exotic answers were tried. In 1909 Camus, at the Vaccine Insti-tute in Paris, following a line of work going back to 1881, dried vaccine pulp by placing it in an evacuated chamber along with an open-topped vessel containing concentrated sulphuric acid, which removed the last traces of moisture from the atmosphere.[36] Variants of this air-dried method, starting with vaccine lymph from cows or buffaloes, were used to produce vaccines which could stand up to tropical temperatures, but the batches were often found to be contaminated with bacteria, and the method was not suitable for large-scale production. Freeze-dried vacuum-packed vaccine turned out to be the answer to both problems, and also to be even better at coping with heat. Freeze-drying of bio-logical material – that is, the rapid freezing of the material followed by rapid evaporation of the ice in a vacuum – was started in America in 1908.[37] It was first used for smallpox vaccine at the end of the 1914–18 war, when the Vaccine Institute in Paris produced a vacuum-packed product that was used successfully in the French tropical colonies in Africa and in French Guiana in South America. Between 1920 and 1940 around 10 million doses of this vaccine were sent annually to the French colonies in Africa; this continued after the war, with the result that when the eradication programme began in the 1960s smallpox was found to be less of a problem in these colonies than in other parts of western Africa.[38]

Other laboratories adopted the French technique. In 1948 the newly established World Health Organisation (WHO) reported approvingly on the Paris vaccine, and it was soon being produced around the world – in Michigan, New York, Peru, Vienna, Copenhagen and Indone-sia. At the Lister Institute in England, Leslie Collier thought out the next step. The phenol, which had been introduced to destroy bacteria, was giving problems when the vaccine was freeze-dried, because when it was concentrated it damaged the virus. Collier's answer was to add peptone (soluble half-digested protein), which prevented the damage and helped preserve the potency of the vaccine:

> *Vaccine thus prepared consistently maintained its original potency for at least three months at 37°C; in later experiments, batches stored at the high temperature of 45°C still gave 100% successful primary vaccina-tion after four years.*[39]

The material that was freeze-dried did not contain glycerine (which would have acted as an anti-freeze) but, suspended in a 40 per cent solution of glycerine when it was about to be used, this was the best form of vaccine for the planned eradication programme.

By 1967, when the second and intensified phase of the global eradi-cation programme started, calf vaccine had been thoroughly tried and tested and could conveniently be produced in developing countries. A world-wide survey of the sources of vaccine in laboratories producing freeze-dried vaccine showed that thirty-nine used calves, twelve used sheep and six used water buffaloes, while three in the Americas had started to use membranes of developing chicks, and three in Europe had started to use tissue cultures – cells of particular tissues grown in glass vessels on sterile artificial media.[40] The advantage of these new methods is that they make it possible to produce a vaccine completely free of bacteria – something that is almost impossible starting with the scarified side of a calf. But although the new methods were already well established for other purposes – for making vaccines against yellow fever and polio, for example – neither was widely adopted for making vaccines against smallpox. At a time when nearly a third of the world's population was still living in countries in which smallpox was endemic, it was more important to have adequate quantities of tolerable vaccine than possibly inadequate quantities of ideal vaccine. Seeing how many pocks a given amount of vaccine would produce on the membranes of chick embryos was, though, already the standard way of assessing vaccine potency.

But if, until very recently, the smallpox vaccine makers were conservative, the Soviet military scientists were not. During the Cold War, less than fifty miles from Moscow, near the great monastery of Zagorsk, they perverted the chick membrane technique to grow vast amounts of variola virus, for biological weapons, in fertilised hens' eggs. Hundreds of thousands of eggs could be sent every month to the Centre for Virology from the state-run collective farms without attracting attention.[41] Calves would have been less productive and more conspicuous.

13

Sorting out the viruses

Especially if the dog was rabid, the virus must be drawn out
with a cupping glass

Aulus Cornelius Celsus, 1st century AD

Celsus was a Roman, and the quotation about the rabid dog is translated from the Latin, but he did use the word '*virus*'. He did not, though, mean what we mean by that word. In Latin, *virus* can mean a slimy liquid, a poison, or an offensive odour or taste.

It's not clear when the word entered the English language. By the beginning of the seventeenth century it was being used to mean poison or venom – hence the word 'virulent' – and during the next two centuries, as ideas of infection became established, it came to mean an infectious agent. In the second half of the nineteenth century a good deal of effort was put into identifying such agents. It gradually became clear that, though most could be removed from the fluid containing them by using suitably fine filters of paper or unglazed clay, others passed through these filters – they were 'filterable'. And there was another striking difference between the two classes of agent. Those that could be removed by filters were visible under the microscopes of that period; those that were filterable were too small to be visible. By the end of the century some investigators were restricting the term virus to the filterable class and were distinguishing between the 'filterable viruses' and the non-filterable infectious agents – predominantly, though not exclusively, bacteria.

So what sort of thing was it that could pass through a filter, be invisible under the microscope and cause disease? Oddly, a crucial part of the answer came not from medical bacteriologists investigating rabies or foot-and-mouth disease, but from a study of a disease of plants – possibly because work on bacterial diseases of humans and animals by Pasteur and Koch had been so successful that it was difficult to think beyond their concepts;[1] in contrast, little attention had been paid to infectious diseases of plants, most of which were then thought to be caused by fungi.

In 1898 Martinus Beijerinck, a Dutchman working in Delft, made an extraordinarily prescient suggestion about the nature of viruses.[2] Studying the agent responsible for the mosaic disease of tobacco plants, he found that the sap of infected plants remained infectious even after it had been passed through a fine filter of unglazed porcelain. This had been noticed six years earlier by Dmitri Ivanovsky, in Russia, who had suggested that the active agent might simply be a soluble poisonous substance – diphtheria toxin had been discovered four years before Ivanovsky's experiments. But Beijerinck showed that the agent could not just be a poisonous substance, because it increased in amount when injected into healthy tobacco plants. And, surprisingly and significantly, he showed that, unlike bacteria, it increased in amount only in the presence of tissues of the host plant that were growing and whose cells were dividing. Because it was filterable he thought (wrongly, as it turned out) that it must be a soluble molecule. It seemed to him inconceivable that a molecule could reproduce itself, so the fact that it multiplied only when growing tissues were present led him to suggest that it could be reproduced only when it was *incorporated into the living protoplasm of the cell, into whose reproduction it is, in a manner of speaking, passively drawn*.

In the twentieth century, this inability to reproduce except by using the reproductive machinery of the host proved to be the defining feature of viruses. And we now know why. Unlike all other living organisms, viruses carry the genetic *information* necessary for their reproduction but they lack at least some of the machinery needed to make use of that information. Put more concretely, they have the DNA (or very similar RNA) containing the necessary genes, but they lack some of the enzymes needed to replicate the genetic material and to use the information in it to synthesise proteins. Indeed it is entirely a matter of definition whether or not viruses are regarded as living organisms. Unfortunately, choosing to regard them as non-living does not reduce their potency as agents of disease.

Because viruses cannot reproduce without machinery provided by their hosts, they are necessarily parasitic, and their hosts include bacteria, blue-green algae, fungi, plants, invertebrates (including insects) and vertebrates (including humans). The viruses responsible for smallpox and cowpox belong to a small group known as the orthopox viruses, which, as their name suggests, all cause diseases that involve eruptions of the skin – pocks. They are closely related genetically, and they all depend on mammalian hosts, though some are choosier than others.[3] Just as the virus that causes smallpox (variola) is found naturally only in humans, there are other orthopox viruses that seem to be specific, or almost specific, for camels or racoons or mice or gerbils. But the virus responsible for cowpox can also live and reproduce and be transmitted from one individual to another in a variety of species, including humans, cats, rats (probably the virus's natural reservoir), elephants, okapis and rhinoceroses. And the virus responsible for monkeypox, which in the wild is found only in the tropical rainforests of central and western Africa, can live, not only in monkeys, apes and humans, but also in anteaters, squirrels, rats, rabbits and prairie dogs. (Curiously, the virus that causes chickenpox is not a member of the orthopox family but is related to the herpes virus responsible for cold sores. It does not infect chickens.)

The orthopox viruses are among the biggest of all viruses – indeed they are just big enough to have been noticed as minute particles under the light microscope in 1886, when they were wrongly identified as bacterial spores.[4] Under the electron microscope they appear roughly brick-shaped. Their genomes, which consist of two strands of DNA, are (by virus standards) very large. Orthopox viruses also contain many proteins: of these, some are structural, some are enzymes involved in the virus's reproduction,* some cause the surface membrane of the virus to fuse with that of the host cell, so allow- ing the virus to infect cells, and some have the role of subverting the defence mechanisms of the host. One of these defence mechanisms involves a small group of proteins known as *interferons*, which act in a complicated fashion to interfere with viral reproduction; and pox viruses contain at least four proteins which block the actions of these interferons in a variety of ways. Evolution over a very long period has made these viruses horribly sophisticated.[5] There is, though, one feature that makes them vulnerable: the virus does not change in the awkward way that, say, flu viruses change, making immunity to this season's flu ineffective against next season's.

The relationships between the different orthopox viruses have been interesting ever since Jenner infected the eight-year-old James Phipps with material from the cowpox pustule on Sarah Nelmes's hand and showed that he became unresponsive to a subsequent inoculation with material from a smallpox pustule. We now know that this is because many of the proteins of the two different viruses are so similar that the elaborate and specific defensive measures set in motion in the human body by the presence of one virus are also effective against the other. These specific defensive measures – the so-called 'immune response' – are of two kinds, though both depend on the ability of certain white cells in the blood and in the lymph glands to recognise the presence of foreign molecules (for example, the particular proteins of bacteria or viruses), and to react in characteristic ways.

Cells known as *B-cells* – because cells of this type were first discov- ered in an organ in birds known as the *bursa* of Fabricius, and later

---

*Most viruses with genomes consisting of DNA rely on the nucleus of the host cell to pro- vide the enzymes necessary for replicating the viral DNA and for transcribing the informa- tion in that DNA into messenger RNA – the molecule that directly controls the synthesis of new protein. The orthopox viruses (and some other members of a wider group of pox viruses) are peculiar in providing the DNA replicating and transcribing enzymes them- selves; they can therefore, exceptionally among DNA viruses, reproduce themselves in the cytoplasm of the host cell. Like all viruses, they do depend on the host cell for the machinery that synthesises protein and for a supply of energy.

(conveniently for the nomenclature) in the *b*one marrow of mammals – react by producing antibodies. These are soluble proteins that circu-late in the blood and combine specifically with the recognised foreign molecules, forming complexes that are then attacked and destroyed by a battery of proteins that immunologists curiously call 'complement'. At the first appearance in the body of foreign molecules of a particu-lar kind, only very few of the B-cells will recognise them; but some of those that do will then divide repeatedly to form a clone of cells – immunologists call them 'memory cells' – which persist in the body so that the next time foreign molecules of that kind appear the immune response is much more vigorous. This type of immune response has been understood, in outline, for more than a century, but it is more effective against bacteria than against many viruses – simply because viruses spend most of their time inside the host's cells, where they are inaccessible to circulating antibodies. Despite this limitation, these cir-culating antibodies are crucial in providing immunity to viral infection because, by attacking viruses before they enter the host's cells (which viruses need to do before they can multiply), they greatly hinder viral reproduction.

In the course of the last century it was gradually realised that there is another arm to the immune system, which can cope with foreign mol-ecules that are inside cells, and which is crucially important in deal-ing with cells that contain viruses. There is a continual breakdown of proteins in living cells, and in most vertebrate cells there is elaborate machinery for presenting the resulting fragments on the surface of the cell, bound to certain 'presenting molecules' in the cell membrane. If the fragments presented on the surface come from foreign molecules (for example from a virus living in the cell) they are recognised as foreign, not by B-cells but by other cells circulating in the blood known as *killer T-cells* – 'T' because they originate in the *t*hymus gland, and 'killer' because assassination is their job. They have, anchored in their outer membrane, receptors of which the exposed part resembles the business end of an antibody molecule. If the receptor recognises, and binds to, a foreign molecule presented on the surface of a host cell, a signal is sent to the T-cell, which then kills that host cell using two quite different methods of attack. The first method is for the T-cell to release a protein called *perforin*, which, as its name suggests, makes holes in (perforates) the membrane of the infected host cell. The second method involves triggering a remarkable self-destruct mechanism present in many animal cells and discovered about thirty years ago.[6] A feature of this cell suicide — its discoverers called it *apoptosis* from the Greek for falling leaves — is that the cell shrinks rapidly and is engulfed by scavenger cells without releasing any of its contents; so a virus-infected cell can be eliminated without releasing any viruses. The use of two separate methods of attack may seem like overkill but, as we have seen, viruses have plenty of spanners to throw into the host's complicated defensive machinery, so an attack on more than one front is worthwhile. Like B-cells, T-cells can generate 'memory cells' so that the second exposure to foreign molecules of a particular kind causes a much more vigorous response.

Cross-protection similar to that demonstrated by Jenner between cowpox and smallpox occurs, though not always as strongly, between many pairs of orthopox viruses, and studies of cross-protection became the classical way of showing the relationships between the different viruses in this group. Nowadays much more detailed information can be obtained by looking at the DNAs in the different viruses. These relationships are relevant to two unsolved puzzles, one of them quite unexpected.

The first puzzle is: how could the variola virus (the cause of smallpox) have evolved? Because it has no animal reservoir, and causes a disease that is infectious for only a few weeks and is followed by a lasting immunity if the patient recovers, the virus can survive only in large human communities. In its present form, then, unless it had an alternative animal host that has died out, it cannot have existed before about 10,000 years ago (when settled agriculture began), and it must have appeared, at the latest, by 3,000 years ago if it really was smallpox that attacked Ramses V.

One possibility is that a virus that was able to survive in *small* human communities – because it produced a more lasting infection or less last/ ing immunity – changed its character to resemble the variola virus we know. That hypothesis would be more attractive if a virus with the required characteristics could be found among the hunter/gatherer populations that still exist.

Another possibility is that a virus that could live in either humans or an alternative host – perhaps a domestic animal – changed so that it became specific for humans. Possible candidates among the known orthopox viruses are those that cause cowpox and monkeypox. Until very recently, the virus causing monkeypox seemed the strongest can/ didate, since it not only infects monkeys, humans, squirrels and anteat/ ers, but produces a disease in humans very like smallpox, with a case fatality that is low in adults but can be as high as 15 per cent in chil/ dren under five years old.[7] Judging by its behaviour in small villages in dense tropical rainforest in Zaïre (now the Democratic Republic of Congo), though, monkeypox spreads so feebly from person to person that it would be unlikely to persist in human communities unless it was continually reintroduced from an animal reservoir.[8] Currently, of course, monkeypox in central and western Africa is cut off from the Nile valley – the nearest site of early large agricultural settlements – by the Sahara desert, but between about 8000 BC and 2000 BC much of what is now that desert was savannah, with evidence, both from skel/ etons and from rock paintings, of giraffes and elephants and people.[9] Whether it also contained enough monkeys, rodents or other suscepti/ ble animals to carry the disease across the gap between the central Afri/ can rainforest and the Nile valley is an open question.

In any event, the notion that the variola virus evolved in the last 10,000 years from the virus that causes monkeypox seems improbable in the light of recent information about the DNAs in the two viruses.[10] Both genomes have now been sequenced in their entirety, along with the genomes of the other orthopox viruses, and it is clear that – contrary to earlier views based on the sequencing of only parts of the genomes – within the orthopox group, the virus responsible for monkeypox is the species that least resembles variola.

On the other hand, an old Arab belief that smallpox originally came from camels now seems nearer the mark.[11] It had been largely dismissed because camelpox rarely if ever attacks people, but Caroline Gubser and Geoffrey Smith, at Oxford and Imperial College, London, have shown that the camelpox genome is closer to that of variola than is the genome of any other orthopox virus for which we have DNA sequence data.[12] It seems, too, that both viruses possess some stretches of DNA unlike anything in the other – compatible with both having evolved from a common ancestor, possibly a virus infecting rodents. And there are other resemblances between the two viruses. Both produce similar small white pocks when grown on the outer membrane of a developing hen's egg, both will grow only within the same restricted range of temperatures, and both are unusual among orthopox viruses in being unable to grow in rabbit skin. And though camelpox does not attack people and smallpox does not attack camels, back in 1975 Derrick Baxby in Liverpool and his collaborators in Teheran showed that inoculation of camels with a strain of variola, caused only a small local response, but protected the camels from camelpox.[13] (Whether inoculation of people with camelpox would protect them against smallpox is a question that we hope there is never an opportunity to answer.)

Now, the unexpected puzzle.

There can be little doubt that many of the early vaccinations were inoculations of cowpox. We even know the name of the cow that is supposed to have infected Sarah Nelmes: she was called Blossom and her portrait hangs in the Jenner Museum at Berkeley. And of course the word vaccination (until Pasteur honoured Jenner by widening its meaning to mean deliberate immunisation against any infection) implied that the inoculated material came from a cow – either directly, or indirectly through the arm-to-arm method. So it was natural to assume that the agent being inserted into the skin when people were vaccinated was generally cowpox; and the cowpox virus in the vaccine acquired the alternative name of *vaccinia*. In 1939 though, Allan Downie at the London Hospital showed that the biological properties of the vaccinia viruses in three different com-

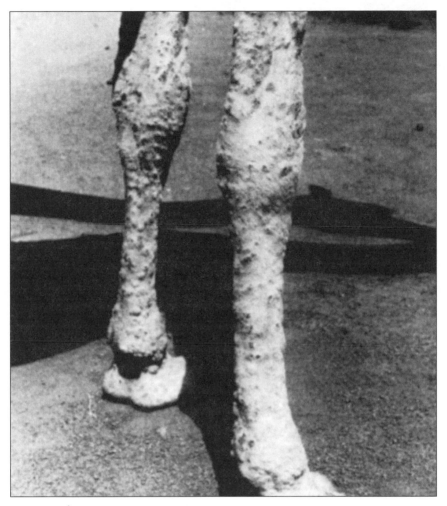

29. Camelpox

mercial strains of vaccine differed markedly from the properties of the viruses obtained from spontaneous cases of cowpox.[14] For example, when cowpox virus was grown on the outer membranes of developing hens' eggs, it produced bright red haemorrhagic pocks, quite distinct from those produced by the various strains of vaccinia virus, and distinct too from those produced by the variola virus. More recently, analysis of the DNAs of different strains of cowpox, vaccinia and variola viruses confirms that, though all share many features characteristic of orthopox viruses, the three species are quite distinct.[15] What then is the origin of the vaccinia virus?

The short answer to this question is that nobody knows. At some time between Jenner's experiments in 1796 and Downie's in the 1930s, for unknown reasons and without being aware of it, those engaged in making vaccines against smallpox seem to have replaced the traditional cowpox by another orthopox virus which shares the same crucial property: when inserted into human skin it confers immunity to smallpox while causing a response that is usually mild and is not accompanied by any risk of catching or spreading smallpox.

A longer answer may not be more useful but is more interesting. Cows were, of course, not the only source of smallpox vaccine. We have seen how Jenner himself, the Yorkshireman John Loy and the Italian Louis Sacco used material that came, directly or indirectly, from lesions on horses; and Jenner even supplied the National Vaccine Establishment with equine material. One possibility then, suggested by Baxby, is that the vaccinia virus is derived from horsepox.[16] Unfortunately, as Baxby has pointed out, since horsepox seems to have died out in the course of the twentieth century, there is now no way of comparing its DNA with that of vaccinia. It is interesting, though, that the Ankara strain of vaccinia, a modified version of which is currently of great interest as a basis for safer vaccines, is known to have originated in a horse in Turkey in the last century.[17]

For a long time, another hypothesis was that vaccinia arose by transformation of the variola virus itself. Though we now know that smallpox and cowpox are caused by distinct viruses, Jenner did not know that; and he not only called the mild disease produced by inoculating humans with cowpox 'variolae vaccinae' (smallpox of the cow), but, in a letter, he even referred to the vaccine as 'the smallpox in a purer form'.[18] If smallpox could be made milder and uninfectious by passage — immunologists pronounce it Frenchwise to rhyme with massage — through a cow, it seemed worth inoculating cows with variola virus and seeing whether the contents of any pustules produced could be used to inoculate humans against smallpox. And, as we have seen, this was not just of theoretical interest; if it worked, vaccination need never be held up for lack of the elusive cowpox.

The experiments that had started with Gassner of Günzburg in 1801 continued until the 1960s with startlingly contradictory results. In the latter half of the nineteenth century two groups developed, the *unitarians*, who believed that transformation was possible, and the *dual-ists*, who regarded the two species as separate with no possibility of transformation. In the twentieth century, as experimental techniques improved, the balance swung away from transformation, which anyway began to seem increasingly unlikely as more was learnt about the indi-vidual viruses.

A possible explanation of the occasional experiments in which transformation was claimed was suggested by Albert Herrlich and his collaborators in Munich and Tübingen in 1963.[19] They suspected that such experiments might have been the result of using matter from smallpox pustules of patients who had been vaccinated only a short time before they became ill with smallpox. If the matter injected into a cow contained both variola and vaccinia viruses, the vaccinia virus would probably flourish, while the variola might, at best, just survive. And they point out that throat washings taken from a patient in Hei-delberg who had recently been vaccinated contained both variola and vaccinia.

Yet another hypothesis was that vaccinia arose as a hybrid between variola and cowpox. There is no doubt that when host cells contain two different species or two different strains of orthopox virus, bits of DNA can be interchanged between the different forms, leading to the formation of hybrids; and in the long history of vaccine production there must have been many occasions when hybrid formation could have occurred. But comparison of the DNAs of variola, vaccinia and cowpox does not support the theory.[20]

You might think that a good way of discovering the nature of the vaccinia virus would be to look at the history of the various famous strains of vaccinia that were used to eradicate smallpox in the course of the last century. Take the *Lister* strain, for example, the strain that played an important part in the eradication programme and that the British government recently ordered 3 million doses of. Surely the venerable Lister Institute would know precisely how that strain had been produced. In fact there is an undocumented tradition that it was originally isolated in the Vaccine Institute in Cologne from a Prussian soldier suffering from smallpox in the Franco–Prussian war in 1870. But it was received in London in 1907 in the form of calf lymph;[21] and in view of the unlikelihood of transformation, even if the tradition is true it is extremely unlikely that the effective virus was an attenuated form of variola. It may have been the virus in the Cologne Institute's own vaccine strain, whatever that was. Or take the *New York City Board of Health* strain. It was first made in the New York City Department of Health Laboratories in 1876, using a 'seed virus' brought over from England. It was later distributed to many other laboratories, where it acquired different names and different biological properties when it was transferred from animal to animal in different ways.[22] Finally, take the *Temple of Heaven* strain, the strain used for the eradication programme in China. 'In 1926', Arita tells us, 'pus from a smallpox patient was passed 3 times in monkeys, then 5 times in rabbits (skin/testes), 3 times in calf-skin, 1–2 times in rabbit skin and a further 1–3 times in calf skin.'[23] And he adds laconically, 'Contamination with vaccinia virus probably occurred during these passages.'

The close resemblance between the DNAs of these various strains of vaccinia – with their different bizarre provenances with 'eye of newt and toe of frog' overtones – can be accounted for by selection of particular strains with desirable properties, and does not require some sort of convergent *transformation* of originally different viruses. This does not mean that mutations never occur. They presumably account for the slight differences between different strains of several of the orthopox viruses and, more dramatically, for the appearance of variola minor at the end of the nineteenth century. Recent comparisons of the genomes of variola minor viruses from different sources show that there is considerable variation, with the Somalian version being closer to variola major[24] – a finding that may explain longstanding reports of strains of variola with effects intermediate between those of variola major and variola minor.

Fortunately, not knowing the answer to the vaccinia problem does not affect the efficacy of the vaccines. In eradicating smallpox, physicians were often, and not for the first time, using successfully remedies they did not fully understand.

14

Eradication: the beginning of the end

Like 'engrafting' and 'inoculating', 'eradicating' is in origin a gardening term — 'rooting out'. That smallpox has been rooted out from all parts of the world is, by any standards, a staggering achievement — both medically and politically one of the most remarkable achievements of the twentieth century. And the scale of the problem was greater than anyone imagined at the start. In 1958, when the Russians suggested that the World Health Assembly should consider the global eradication of smallpox, they quoted the latest annual figure for cases of smallpox worldwide at over 132,000.[1] Because of under-reporting this turned out to be a gross underestimate — Donald Henderson, the epidemiologist who was to play a crucial part in the eradication of smallpox, suggests the annual number of cases in the early 1950s might have been as high as 50 million, dropping to around 10 to 15 million by 1967.[2]

At the beginning of the twentieth century, most of Australasia and most of the Scandinavian countries were already free of endemic smallpox, and by the middle of the century they were to be joined by all the rest of Europe. But no place was proof against hazards of travel or of war. In 1900 smallpox had been a serious problem in Mediterranean countries with seaboards exposing them to infection from North Africa;[3] outbreaks in London and in Belgium in 1901–3 were thought to have originated in France. In the 1914–18 war Russian Poland had bad outbreaks, and in the two years after the war Russian prisoners, and returning soldiers, caused 1,500 smallpox deaths even in well-vaccinated Germany, while 28,000 died in Italy and 14,000 in Portugal.[4]

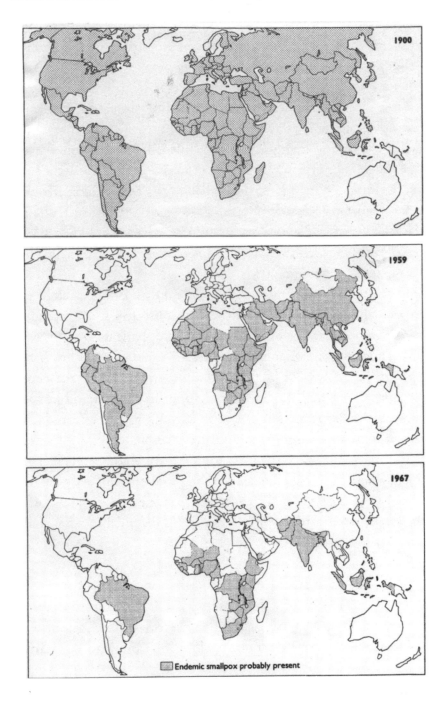

30. Maps showing countries where smallpox was endemic in 1900, 1959 and 1967

In the United Kingdom, variola major was no longer endemic after 1905, but it was frequently brought into the country causing outbreaks in most years until 1929 and occasionally later – in 1946, with the repatriation of troops after the Second World War, it was imported on fifteen separate occasions giving rise to fifty-six cases.[5] The mild variola minor became established in the United Kingdom in 1919, peaked in 1927 and then gradually declined; by 1934 it too was no longer endemic.[6] Surprisingly, despite criticism in the British medical press, the United Kingdom, unlike other advanced industrial nations, did not require travellers from overseas to carry a valid international vaccination certificate until 1963.[7] (But perhaps we should not be surprised. Unlike the USA, Canada and Australia, Britain currently has no health tests for immigrants.)

In North America, endemic variola major had disappeared by the late 1920s, and from then to the late 1940s the disease was almost entirely variola minor. Because this disease was mild, patients were likely to go about infecting others, and because the mortality was very low it was treated less seriously both by the health authorities and by the population at risk. An outbreak of variola minor in Switzerland in 1921 dragged on for six years, and is said to account for the extraordinary Swiss response to a suggestion, at an International Sanitary Conference of the League of Nations in 1926, that smallpox be made internationally notifiable. Smallpox, according to the Swiss delegate, had 'no place in an international convention. It is not a pestilential disease in the proper sense of the term: it is, in effect, a disease that exists everywhere. There is probably not a single country of which it can be said that there are no cases of smallpox.'[8]

It was variola major that was endemic in eastern and south-eastern Asia. Figures for China are vague, but smallpox seems to have been widespread (and vaccination scarce) all through the first half of the century until a national eradication campaign was started in 1950. Vaccination did not reach Tibet at all until 1940; Japan, by contrast, produced calf vaccine efficiently, but suffered from frequent bouts of imported smallpox; Korea, which exported some

of those cases, suffered particularly badly in the 1951 war, but managed to get rid of smallpox altogether three years later; the Philippines,
after much trouble early in the century, were clear by 1931; Thailand
and Burma were not clear until the 1960s. Edmund Hillary, in his
mission to improve the life of the sherpas who had done so much for
him, organised hundreds of vaccinations in Nepal in an epidemic in
1963.[9] The Indian subcontinent had the doubtful distinction of being
the major focus of smallpox in the twentieth century until eradication
was achieved there in 1975.

In North Africa, vaccination campaigns run by the colonial powers
dealt effectively with smallpox everywhere except Morocco before the
start of the Second World War, though the disturbed times of war
brought outbreaks that rumbled on for the next ten years. Further south
the colonial powers do not seem to have managed so well, with endemic smallpox surviving in most of western, central and eastern Africa
until the end of the 1960s; the chain of infection finally ended in 1976
and 1977 in Ethiopia and Somalia. Further south still, during the First
World War Madagascar had become the first African country to be
free of smallpox. There was endemic smallpox, often predominantly
variola minor, in all the other southern African countries throughout
the first half of the century.

Planning to get rid of smallpox altogether was not just philanthropy;
it was in everyone's interest, because there would then be no danger of
imported infections and no need for general vaccination. In 1950 the
Pan American Sanitary (later Health) Organisation, prompted by an
outbreak of smallpox in New York in 1947 caused by a visitor coming
by bus from Mexico, started a scheme for eradicating smallpox from
the Americas. The scheme got off to a slow start, but by the end of
1958 endemic smallpox had been eradicated or nearly eradicated in all
but six countries – Brazil (where it was mainly variola minor), Colombia and Ecuador with many cases every year; Argentina, Bolivia and
Paraguay with a few.

Ambitions broadened in 1953 when the first Director-General
of the WHO, the Canadian Dr Brock Chisholm, proposed at the

Sixth World Health Assembly that the organisation should undertake a world-wide programme for the eradication of smallpox. But the
delegates were not enthusiastic: the problem was really 'regional', it
was 'vast and complicated', 'might prove uneconomical', was not suitable at this time. So the matter was deferred to the next World Health
Assembly in 1954, and deferred again to 1955, when it was dropped
– eclipsed for a further four years by a proposal to adopt a programme
for the global eradication of malaria, which the delegates felt should be
given the highest priority.

Since vaccination against smallpox had been outstandingly successful, and the eradication of malaria (in the absence of any method of
immunisation) would have to depend on getting rid of the mosquitoes
that are its intermediate hosts, it may seem odd that malaria should have
been the preferred target. But there was a tradition of earlier campaigns
against mosquito-borne diseases going back to the beginning of the
century, encouraged by the discovery of the insecticide DDT in the
early 1940s. When mosquitoes started to become resistant to DDT,
the worried Pan American Sanitary Conference declared regional malaria eradication to be an emergency need and authorised a special fund
of $100,000 for administrative expenses. In 1955 the World Health
Assembly agreed that eradication should be a world-wide policy, and
that the WHO should aim to achieve it 'before the potential danger
of a development of resistance to insecticides ... materialises'. In the
course of the next four years annual expenditure on the eradication of
malaria, from the WHO and from other sources, increased sixfold;
and so did the number of WHO staff concerned with the eradication
programme. That high level was kept up for a further ten years.[10]

This colossal effort – nearly $14 million annually, and about 500
staff – produced substantial results; but by 1969 it had become clear
that, even with this effort, global eradication of malaria was not going
to be achieved, and in 1973 the programme ended. Its relevance to
smallpox eradication is not only that it delayed the start of a global
programme for smallpox, but also that when such a programme was
started, in 1959, it still had to compete for resources with the vastly

more expensive malaria programme; and it made many people sceptical of the whole concept of global eradication.

The driving force behind the 1959 smallpox eradication programme was Viktor Zhdanov, Academician and Deputy Minister of Health of the USSR. Unusually among ministers dealing with medical and scientific matters, he was an expert in the field with which he was concerned. And, thanks to a change in policy by Khrushchev, he was the first Russian representative at the Assembly since Stalin had ended Russian participation in United Nations agencies ten years earlier, and was therefore particularly welcome. He started tactfully by reminding his audience that the American President Thomas Jefferson, writing to Jenner in 1806, had predicted that Jenner's discovery would ensure that 'in the future the peoples of the world will learn about this disgusting smallpox disease only from ancient traditions'. Zhdanov then went on to argue that it was now scientifically feasible, socially desirable and economically worthwhile to attempt to eradicate smallpox world-wide. A subsequent resolution of the World Health Assembly put the point squarely: 'funds [currently] devoted to the control of and vaccination against smallpox throughout the world exceed those necessary for the eradication of smallpox in its endemic foci'.[11]

Zhdanov therefore proposed 'a programme for the eradication of smallpox over a period of three to five years'. To achieve that, the WHO should plan: to place orders with national biological firms for the preparation of adequate quantities of vaccine; to train local native smallpox vaccinators; and to find local sources to supplement WHO funds for buying the vaccine and paying the vaccinating teams. After the first series of vaccinations, 'provision should be made for additional vaccination … in the foci where the disease still occurred'. In tropical or subtropical countries, it would be best to vaccinate in the cool season. Finally (and remarkably, fifty years after Millard), he referred to the Leicester system – i.e., the prompt identification, notification and isolation of new cases, and the quarantine and surveillance of contacts – and recommended that it 'should be used as fully as possible' because 'together with vaccination it will greatly accelerate the eradication of smallpox'.[12] If all this were done, he prophesied, smallpox would be 'practically eradicated within five years', and 'its complete eradication throughout the world' would be 'achieved within the next ten years'. (He did not mention, and perhaps did not know, about the smallpox weapons factory which had been running for the last ten years at Zagorsk.[13])

Despite some reservations, the delegates were persuaded that a global eradication programme was timely, and they asked the Direc-tor-General of the WHO, then Marcelino Candau of Brazil, to look at the proposals in more detail. After the meeting the WHO accepted gifts of 25 million doses of freeze-dried vaccine from the USSR and 2 million doses of glycerinated vaccine from Cuba; and later in the year its Biological Standardization Committee established internation-al standards for vaccines. A resolution of the Twelfth World Health Assembly in 1959 at last committed the WHO to a global smallpox-eradication programme.[14]

The programme began with three weaknesses – one strategic, one financial and one political. The strategic weakness was that the reso-lution concentrated wholly on the vaccination or revaccination of at least 80 per cent of the population in each country, and there was no mention of the importance of using the Leicester system as a backup to mass vaccination. The financial weakness was that despite a huge gap between the funds likely to be available and the estimated cost of vacci-nating or revaccinating not far short of a billion people in the endemic areas, there was no clear policy for bridging that gap; the responsibility for meeting costs would rest with individual governments.

The political weakness was that the People's Republic of China was not then a member of the United Nations, so it was not clear how the eradication of smallpox in China could be achieved – or reliably known about, if it were achieved. In fact, as became clear years later, this turned out not to be a problem, because of the success of three mass campaigns of compulsory vaccination and revaccination organised by the Chinese government itself in the 1950s and early 1960s.[15] Between 1950 and 1963 they issued 1.8 billion doses of vaccine, nearly all of it calf vaccine of the Temple of Heaven strain. In 1959, the year of the start of the global campaign, only the province of Yunnan reported cases of smallpox, though there were one or two cases in other regions for the next six years. Most of the last cases, all in remote mountainous parts of northern China, were thought to be the result of variolation by practitioners of traditional Chinese medicine to whom people turned for help when supplies of vaccine were not available.[16] Although small-pox had disappeared from this part of China in the 1950s, the variola-tors are said to have kept stocks of the smallpox virus mixed with honey in sealed jars, renewing the stocks annually by inoculating children.

Despite the weaknesses of the WHO programme, there was con-siderable progress. In 1959 more than 1.7 billion people – 59 per cent of the world's population – were living in countries where smallpox was endemic. By 1966 the proportion was down to 31 per cent, but the rise in world population brought the total up to more than a billion people.[17] Global eradication was still a long way off.

In 1963 it was suggested that the WHO should create a separate budget for smallpox eradication, financed by member states, but the Director-General, already worried by the expense and limited suc-cess of the malaria programme, was reluctant to increase the WHO's responsibility for another programme that might not succeed. It was generally assumed that eradication required the vaccination of at least 80 per cent of the population, and this would not be achievable in many parts of the world.

Two years later, matters started to look more hopeful. Karel Raska, the distinguished Czech epidemiologist who had become Director of Communicable Diseases at the WHO, persuaded the Director-General to set up a separate Smallpox Eradication Unit – consisting at first solely of the Japanese virologist Isao Arita and a secretary. Then technical improvements – jet injectors and bifurcated needles – made the process of vaccination far cheaper, easier and more effective. At the same time there was a sharp increase in the USA's interest in eradica-tion. In 1964, International Cooperation Year (celebrating the twen-tieth anniversary of the United Nations), President Lyndon Johnson

31. Smallpox demon being slain by a hero with a bifurcated needle. Poster used in the eradication campaign in India. The actual bifurcated needle was only 5cm long with rather sharper prongs. Its notch held just sufficient vaccine, so there was no waste; it could be sterilised repeatedly by flaming or boiling; and unskilled people could be taught to use it effectively in only twenty minutes. Mass-produced, it could also be cheap; the WHO had 50 million of them.

had announced American intentions to 'expand our efforts to prevent and to control disease on every continent'. The following year he instructed the US delegation to the World Health Assembly to 'pledge American support for an international program to eradicate smallpox completely from the earth within the next decade'.

As a first move, the USA decided to provide technical and financial help for a programme to eradicate smallpox and control measles in eighteen contiguous countries in western and central Africa. This decision came about in an odd way, and marked the arrival on the scene of Donald Henderson. He had originally joined the American Epidemic Intelligence Service as an alternative to two years in the army, but had become absorbed by the problem of smallpox when he studied the responses of smallpox-free countries to imported cases, and the possible complications of smallpox vaccination. By 1965 he was head of the Surveillance Section of the Communicable Disease Center – the famous CDC[18] – in Atlanta, and smallpox dominated his life.

Through Henderson's influence, a plan to eradicate smallpox was hitched on to an existing American measles-vaccination programme in a small group of countries in west Africa. He argued that since there was no prospect of eradicating measles, the wider scheme would make continued American help worthwhile; but this made sense only if the scheme was widened geographically as well, to include Nigeria and Ghana – for without them smallpox would creep back across the borders. As the combined populations of Nigeria and Ghana exceeded 60 million, this obviously made the programme very much more expensive, but in view of the American pledge at the World Health Assembly (and despite the fact that what they had had in mind was assistance in Kenya and South America), the US government promised to provide the extra help. This proved to be a crucial decision.

15

'Annihilation of the smallpox'

T he phrase is Jenner's – he used it in 1801 – but what finally brought it about was an agreement by the World Health Assem-bly in 1966 to start an 'intensified smallpox eradication programme'.[1] The target, although this was reckoned too rash to be put in the official report, was to annihilate smallpox in ten years.[2] The programme was launched early the following year, and the final case of natural small-pox in the world was recorded in October 1977.

The WHO promised annual funding of US$2.4 million. This was about ten times as much as they had allocated for spending on smallpox eradication during 1966, but it was for programmes in more than thirty different countries, and it covered little more than a third of the estimated annual cost of the programmes *excluding* costs that were to be met by the countries themselves. The rest, it was hoped, would be met by contributions from individual countries or other organisations or by gifts of vaccine. In fact, over the course of the ten-year campaign, international contributions added US$64 million.[3] And the interna-tional help was not only in money and equipment; doctors, epidemi-ologists and other health workers from all over the world took part, working out new plans, supporting, training and advising new teams.

At the start of the programme smallpox was still endemic in four widely separated areas: Brazil; the Indonesian archipelago; sub-Saharan Africa; and the great tract of land that includes Afghanistan, Pakistan, India, Bangladesh (then East Pakistan), and Nepal. The WHO was hampered by the chaotic lack of information and gross under-reporting everywhere.[4]

Besides being better funded, the intensified eradication programme had an improved strategy. Mass-vaccination campaigns using, as far as possible, thoroughly tested freeze-dried vaccines were to be followed by what became known as 'surveillance-containment' – a programme to detect, investigate and seek to contain each outbreak of smallpox. This was, of course, just the combination of mass vaccination and Millard's 'Leicester system' that had been advocated, unsuccessfully, by Zhdanov in 1958. When 80 per cent of the population had been successfully vaccinated, the number of smallpox cases would be expected to fall to a level at which individual outbreaks could be dealt with effectively and quickly. In the event, the strategy became more flexible.

This flexibility owed much to the improvisation and enterprise of the American epidemiologist William Foege, who had gone to Africa in 1964 as a medical missionary, living in a simple Nigerian hut and running a clinic, and was now recruited to the programme by Henderson. In eastern Nigeria, field staff for the intensified eradication programme had begun to train in November 1966, but supplies necessary for mass vaccination had not yet arrived. When a missionary reported cases of smallpox in a small village in the Ogoja Province, Foege used his special contact with the network of missionaries to track down other villages with smallpox; and he decided to seize this opportunity to train his teams in 'containment procedures' – vaccinating thoroughly, but in a limited area in the villages and along adjacent roads, so that the scarce vaccine could be used effectively. By the end of four weeks, forty-three cases had been discovered but the outbreak had been stopped. And by the time the supplies for mass vaccination arrived a few months later, eastern Nigeria seemed to be free of smallpox although less than half the population had ever been vaccinated.

Foege's 'detection and containment' approach[5] was supported by some field studies in East and West Pakistan which showed that even in endemic areas, at any one time only a very small proportion of villages were likely to be infected, and that cases tended to be clustered in one part of a town or in adjacent villages.[6]

The work of detection and containment was best carried out during the rainy season (in September and October in western Africa), when the incidence of smallpox was at its lowest. If it was to succeed, the slow routine disease-reporting system – 'passive surveillance' – had to change to a much more vigorous campaign: alerting the public through newspapers and radio, involving other health service personnel, talk-ing to teachers, village chiefs, mail carriers, people in markets, agricul-tural workers and missionaries. It was important not just to isolate the patient and vaccinate the household, but also to determine the chain of infection and vaccinate contacts along it in the village or market or in neighbouring villages. Mass-vaccination campaigns would, of course, continue in areas where smallpox remained endemic.

The outbreak of war in July 1967, when the eastern part of Nigeria declared itself to be the independent republic of Biafra, made mat-ters more difficult but did not immediately stop all vaccination; there are stories of the Nigerian authorities transferring crates of smallpox and measles vaccine to the Biafrans, leaving them in the middle of a bridge during ceasefires.[7] By June 1970, three and a half years after the start of the intensified eradication programme, smallpox was no longer endemic in western Africa.

This dramatic and unexpected success provided strong support for the Foege system – though that success was helped by the low popula-tion density, the small amount of movement in the population and an established tradition of isolating smallpox patients.[8] But whatever the relative contributions of surveillance-containment and mass vaccina-tion may have been, the achievement of eradication in western and central Africa was extremely encouraging for other endemic countries.

Besides the ubiquitous problem of under-reporting, each zone had its own particular problems.[9]

In Brazil there was the enormous area and a population of about 96 million, including tribes in remote reaches of the vast Amazon basin who had little contact with the national authorities. A vigorous vacci-nation campaign was started in November 1966, and within two years more than 19 million people had been vaccinated; but with so much effort being put into vaccination, surveillance had been neglected. The fact that it was variola minor that was prevalent in Brazil also made detection much harder. In July 1967 twenty-one cases of smallpox were reported in the small town of Branquinhas in a *município* that was supposed to have completed its vaccination campaign only three months earlier. Examination of the records showed that the number reported as vaccinated exceeded the total population; but examination of randomly selected households showed that, in the town, the proportion of people actually vaccinated was only about 49 per cent. Even worse, in two areas in the *município* outside the town the proportion was 75 per cent in one and zero in the other. Clearly, records had been falsified, a discovery that led to the setting up of effective assessment teams, and the appointment of a new director, Oswaldo da Silva, a highly efficient organiser who had worked on malaria-eradication programmes. With improving surveillance and a rising rate of vaccination – about 21 million in 1969 and nearly 32 million in 1970 – the number of outbreaks of smallpox fell sharply. The last known case in Brazil was in April 1971.

The 3,000 islands of Indonesia contained 113 million people, more than half of them crowded on to Java.[10] Because the archipelago straddles the equator, there was no season with low infection rates to be exploited as in the eradication campaign in western Africa. Again unlike western Africa, there was no tradition of isolating patients suffering from smallpox; instead there was a custom of taking sick children to visit relatives, so tending to spread the infection. These difficulties were unavoidable, but there was also a further, gratuitous, difficulty. The WHO Regional Director for South East Asia, resident in New Delhi but crucial to any eradication campaign in Indonesia, had taken the line that the campaign was misconceived because its aims were unachievable.

Despite the difficulties, his pessimism was not justified. There had been a long history of attempts to control smallpox in Java. Stamford Raffles had been able to tell Jenner of his efforts to bring vaccination to the island while he was Lieutenant-Governor (and the population of Java was only 6 million); regular arm-to-arm vaccination had been organised in 1856, and an institute for making vaccine had been established (along with a factory for extracting quinine from cinchona bark) in Bandung in the 1890s. By 1937 there were only a handful of cases, and even these were thought to be the result of importations. There are no figures for the period of the Second World War, but shortly afterwards smallpox seems to have come back with a vengeance, spreading southwards from Malaysia, and causing very large outbreaks by 1949.[11]

The views of the Regional Director had hampered preparations for the intensified campaign, but in the autumn of 1967 the WHO nominated a new Regional Director, Herat Gunaratne, and the Indonesian Ministry of Health appointed a new Director-General for the Control and Prevention of Communicable Diseases, Julie Sulianti Saroso.

It was known that the reporting of cases was inadequate, but it was not appreciated how inadequate until a survey was made of the proportion of children under a year old who had pockmarks.* Further investigation revealed that the success rate of primary vaccinations was only 60–80 per cent, and the number of primary vaccinations being given in a year was slightly less than two-thirds of the number of children born. The remedy was to double the number of vaccinators employed, and to concentrate on primary vaccination, using maternal and child health centres and school health clinics since three-quarters of all smallpox cases were in children under fifteen. Small teams would find and contain outbreaks; the WHO would provide trucks, motor-bikes,

---

*The figure was 4.5 per cent. Allowing for the proportion of children with smallpox in their first year who would have died, and the proportion who would have survived the disease without pockmarks, and knowing also the proportion of smallpox patients who were less than a year old, it was possible to calculate from the figure of 4.5 per cent that there must have been at least ten cases of smallpox for every one reported, giving a total of at least 100,000 cases in Java. Later surveys suggested that this estimate should be doubled.

bicycles and refrigerators, and would help with vaccine production.

There were early local successes, particularly in East Java, but it was to take three and a half years to eradicate smallpox from Indonesia, and ultimate success depended not only on great effort but also, here as elsewhere, on the change to an intense surveillance-containment strategy. The teams conducted regular village-by-village searches, and then house-to-house searches to find individual cases. Any cases would be isolated and any contacts vaccinated. By staying overnight in infected villages the vaccinators avoided missing contacts who worked away from the village during the day.

A cost-effective method of searching for outbreaks was invented by an enterprising member of a team in Bandung who was noticeably unenthusiastic about house-to-house searches yet particularly good at finding outbreaks. He used the less obtrusive approach of visiting schools, where he showed photographs of children with smallpox to both children and teachers, and asked them to tell of cases they knew. His success led to the production of the WHO 'smallpox recognition card', which proved enormously useful throughout the rest of the campaign.

An incentive, introduced late in the campaign when the number of cases was very small but finding them was crucial, was to offer a reward – money or a transistor radio – for any reported. The scheme had an immediate success in 1971, when it led to the discovery that reports of cases of smallpox had been suppressed by a medical officer who was afraid that he might be punished for failing to vaccinate successfully. In 1971 there were about 2,000 cases of smallpox in Indonesia. In January 1972 there were only thirty-four, all in a group of villages in West Java; those were the last.[12]

By the end of 1972, then, smallpox had been eradicated from two of the four great areas in which it had been endemic. In sub-Saharan Africa, too, there had been remarkable progress. Variola major had gone, but variola minor, which had been eradicated from South Africa by the end of 1971, had spread north to neighbouring Botswana. Here, in a poorly vaccinated population and with no active surveillance, it spread quickly causing over 1,000 cases in 1972, though only two deaths. A massive vaccination campaign eventually stopped the epi-demic, but only after a succession of outbreaks which continued until November 1973. The later outbreaks affected mainly members of a religious sect that refused vaccination and concealed cases.[13] Variola minor also flourished in the countries in the Horn of Africa; here it would persist until the completion of a more daunting task, the eradi-cation of smallpox from the Indian subcontinent.

Although the Indian subcontinent is one continuous land mass, there was very little movement across the heavily guarded border separating India and (West) Pakistan; and it seems that after 1966 there were no known cases of the transfer of smallpox across it.[14] There was, though, a great deal of movement between Afghanistan and (West) Pakistan, between India and East Pakistan (now Bangladesh), and between India and Nepal.

Afghanistan had all possible handicaps of politics and geography, and in 1963 the WHO-assisted campaign there had got off to a poor start. There was a population of perhaps 12 million, 90 per cent illit-erate, speaking three different languages, and living in a country part mountainous and part desert, with extreme temperatures, few roads, no railways or navigable rivers, a government whose authority was not always recognised, a rudimentary health service, and such strict observ-ance of purdah that women were not permitted to leave their houses to be vaccinated, or to admit male vaccinators.[15] And the risks of infec-tion were increased both by the movements of a large nomadic popu-lation who spent their summers in Afghanistan and their winters in Pakistan, and by the existence of hereditary variolators, particularly but not exclusively in the remote mountains.

A joint WHO–Afghan assessment team produced a damning progress report in 1969. The following year Afghanistan was the source of an epidemic that spread around the Middle East, started by a family from a village near Kabul trekking west on pilgrimage to Mashhad, in Iran, a country that had been officially free of smallpox for seven years. On arrival, the children developed rashes, infecting many pilgrims who in turn distributed smallpox around the whole country and beyond. Over the next two years there was smallpox in Iran, Iraq, Syria and Yugoslavia. But the 2500th anniversary of the Persian Empire was due to be celebrated with great pomp and massed heads of state at Persepolis in October 1971, so Iran tried first to suppress, and then to minimise, any news of smallpox. In 1971–2 thirty-one cases were officially reported, but when Henderson himself went to investigate, he found that over 2,000 cases had been hospitalised, probably about a quarter of the total.[16] Mass vaccination had been tried, but the vaccine was poor, and it was not until the Iranian Deputy Minister of Health appealed to the WHO in November 1971 for freeze-dried vaccine that the epidemic was checked.[17]

The 1969 report had also made very effective recommendations, setting up a combination of five-man vaccination teams, and two-man assessment teams whose job it was to follow up the vaccination work in a tenth of the villages a week or so later; these were headed by Arcot Rangaraj from the Indian Army Medical Service and the Afghani Abdul Mohamad Darmanger. By June 1972, the first round had been completed and 10.5 million people had been successfully vaccinated. Two further rounds, one for children under fifteen and the other for children under five ensured that the proportion of the children unprotected was less than 10 per cent and often less than 5 per cent. The programme also included arrangements for investigating and containing any reported outbreak of smallpox. The remarkable result of this combination of mass vaccination and surveillance-containment was that the last case of endemic smallpox in Afghanistan was in September 1972, though there were a few imported outbreaks from Pakistan in 1973.

—·—

If eradication was bound to be difficult in Afghanistan, the situation in (West) Pakistan at the beginning of the intensified programme looked more promising. Not only was there a more effective government, a more elaborate health service, better roads and airways and a railway system, but in the greater part of the country measures against small-pox were much more sophisticated. Except in a few remote regions variolation had long stopped, and vaccination was well established. As the field studies in East and West Pakistan had shown, given the relatively high level of immunity in the population, it was best to focus on the detection of outbreaks and containment of cases, especially in urban areas and especially in the wet summer months when there was less smallpox and only a few localities were affected. Mass vaccination should concentrate on primary vaccination in the cities. This advice would eventually prove to be sound but, prophets never being recog-nised in their own country, it took some time to be accepted.

It did not help that, in July 1970, the government divided most of West Pakistan into four provinces, which were almost autonomous in matters of health – the North-west Frontier Province, the Punjab, Sind and Baluchistan. Nor did it help that, in March 1971, a civil war erupted in East Pakistan, leading to the creation of the independent state of Bangladesh nine months later.

A difficulty of quite a different kind was that up to 1971 the principal WHO advisers in Pakistan were 'veterans of the success-ful eradication programme in Iran in the early 1960s', a success that had been achieved wholly by mass vaccination, with no surveillance-containment.[18] This fixation on vaccination accounts for what Hend-erson refers to as 'the unnecessary mass vaccination campaign' in a large part of the Punjab in 1969 and 1970. Expensive and beset with problems, it achieved a million primary vaccinations and 22 million revaccinations in a population of 27 million; but it had little effect on the incidence of smallpox.

Surveillance-containment was introduced in the Punjab the fol-lowing year, and by March 1973, in the peak season for transmission, only eight cases were reported in the whole province. The Director of Health Services decided that eradication had been achieved, and virtually stopped the programme. His decision turned out to be premature, for in late October there were cases in Lahore, and by the end of December the number of people in the Punjab who had been attacked by smallpox during the past year appeared to be more than in any year since 1948. This must partly have reflected better detection, but it forced the authorities to take eradication more seriously – as did unflattering comparisons between the performances of the Punjab and of the North-west Frontier Province, where endemic smallpox had already been eliminated. Omer Sulieman, a new adviser from the WHO who had run the successful eradication campaign in the Sudan, wrote a withering report on the arrangements he found and attacked the problem energetically, offering the encouragement of ten rupees for each outbreak reported. Smallpox was eradicated from the Punjab by the end of October 1974. By that time the campaign had also been successful in both Sind and Baluchistan, so the last case in the Punjab was also the last case in Pakistan.

The eradication of smallpox from India was a task of heroic proportions.[19] In 1967 the country had a population of 513 million, and a birth-rate of 25 million. As recently as the 1950s (on the later revised estimates) there had been more than a million deaths from smallpox every year. Four fifths of the population lived in rural areas, and there were reckoned to be more than 5 million villages. Although there were only four towns with populations over 2 million, much of the vast Ganges river plain was densely populated. The people were also extraordinarily mobile – travelling to do business or to attend marriages, funerals, religious meetings, fairs. The huge railway system – India's largest employer – ran nearly 11,000 trains daily and, including road vehicles, the state transport system is said to have carried 10 million passengers a day. What made matters worse was that the favourite time for travelling was in the cool dry season when the smallpox virus was best able to survive. Variolation, though, was no longer a problem, except possibly in some remote areas bordering Pakistan.[20]

For many, the attitude to smallpox was bound up with religion. It

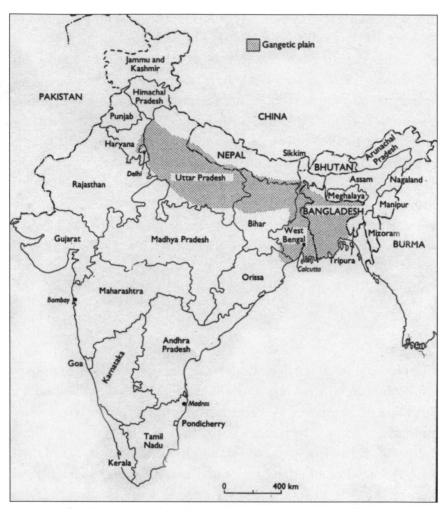

32. Map of India showing the individual states and the Ganges plain

was accepted as an inevitable part of life, and belief in a goddess with power both to inflict and cure smallpox was still widespread, particu-larly in Rajasthan. Sensitive to these feelings, at times the early mass vaccination campaign used the approach 'worship the Goddess and also take a vaccination'[21] – reminiscent perhaps of Cromwell's advice to his troops, 'trust in God and keep your powder dry'.

Shortly after the 1959 decision by the World Health Assembly committing the WHO to a global eradication programme, the Indian

Council for Medical Research had set up pilot projects for mass vaccination in each state; although these were disappointing, the main
scheme was started in 1962, with some national coordination and very
substantial national financial support. The initial aim was to vaccinate
80 per cent of the whole population, but all too often this magic figure
was reached by false reporting or by including repeated vaccinations of
readily available schoolchildren. And the use of heat-sensitive liquid
vaccine meant that there were many failures.[22]

In 1967 more cases of smallpox were reported in India than in any
year since 1958. As in the Punjab, much of the increase must have
reflected more efficient reporting, but coming just after the national
mass-vaccination programme was supposed to have reached its target,
the figures were not encouraging. As in Afghanistan, a joint national–
WHO assessment team was appointed, this time headed by the Swiss
virologist and epidemiologist Nicole Grasset, who had had her baptism
of field experience working for the Red Cross during the Nigerian civil
war and for the WHO smallpox campaign in Indonesia. The team
spent six autumnal weeks visiting four states: Maharashtra and Uttar
Pradesh, both then experiencing epidemics; Punjab (the part still in
India), with a moderate incidence; and Tamil Nadu, with a very low
incidence. The advice was familiar – more emphasis on the detection
and containment of cases, especially during the hot, humid summer
months when there was less smallpox; and vaccination programmes
concentrating on primary vaccination, including the vaccination of
new-born infants. They also recommended that vaccine should be produced and financed centrally, and that the painful and clumsy 'rotary
lancet' should be replaced by the bifurcated needle.

One extra complication on the Indian scene was the number and the
way of life of beggars, wandering from village to village, often spreading smallpox as they went. However ill they might be they refused to be
isolated or confined, as that would mean no income for them or their
families. On one occasion late in the campaign, Nicole Grasset and
William Foege (who had joined the WHO Indian team in 1973) set
up an unconventional solution, housing and feeding infected beggars

and their families, supporting them until they were fit. Foege showed his imaginative approach another time by exploiting his great height (six foot seven) to encourage children to come out and be checked for vaccination scars; he got the village chief to call them to come and see 'the tallest man in the world'.[23] And he was not the only one to improvise. Mary Guinan, a New Yorker from the CDC in Atlanta, faced with the problem of crossing rivers in her travels around Lucknow, accepted the offer of the loan of an elephant. 'Sometimes I could use a camel, sometimes I could get a boat. But if I knew there was critical timing involved ... I would try to arrange in advance for the elephant.'[24]

When the WHO team got to work, after the peak incidence in 1967, the number of cases of smallpox fell sharply in each of the three following years — partly from the normal drop in the number of susceptible subjects after an epidemic. The fall was particularly dramatic in southern India — home to nearly 40 per cent of the Indian population — where the annual incidence of smallpox fell from 42,000 to fewer than 800. In Andhra Pradesh, smallpox cases were mostly found among unvaccinated fishermen who moved seasonally to and from its northern neighbour Orissa. Unvaccinated migrant workers were a problem too in Gujarat, which by 1969, with a population of only 27 million, accounted for a third of all cases in India. Fortunately the epidemic there responded well to a vigorous containment programme.[25]

In Rajasthan, despite the low literacy, the extensive deserts, the few roads and a less developed health service, surveillance-containment was so effective that, starting in August 1973, the state enjoyed a three-month period without smallpox, and subsequent cases were traced to importations. Detection of cases in Rajasthan was greatly helped by the opportunity to question visitors to the many temples dedicated to the smallpox goddess, and by the local habit of hanging branches of the neem tree over the front door of houses containing patients.[26] (The influence of the smallpox goddess was not always so helpful. In one area in Nepal where smallpox spread unusually rapidly, it turned out that the inhabitants were members of a sect that refused vaccination and believed that children with smallpox were possessed by the goddess and should be granted every wish – including the wish to visit relatives.[27])

To the east of Rajasthan, in Uttar Pradesh, with its 91 million people, most of them in dense settlements on the Ganges river plain, the outlook was grim. The mass-vaccination programme had been badly managed, and the number of cases of smallpox rose during 1972 and exploded in the spring of 1973, reaching 5,000 reported cases in May. Earlier that year there had been an extra problem in the heavily infected district of Muzaffarnagar; under the leadership of the WHO, district health staff had set up an effective containment system, but this had hardly got going when the state Director of Health Services ordered a disastrous change of policy from containment to immediate vaccination of the whole population of the district. He also warned that when the mass vaccination had been completed the report of any further cases would result in the transfer of the district health officer to the most unpleasant post in the state.[28] In the first six months of 1973 almost 90 per cent of the reported cases in India were in Bihar, Uttar Pradesh or West Bengal. These three states – all in or near the Ganges river plain – contained less than half of the population of India; they also contained more than half of all the reported smallpox cases in the world.

January 1974 began with a funding crisis, which was fortunately soon resolved when the WHO executive agreed to transfer to the South-East Asia Regional Board – responsible for smallpox eradication in the Indian subcontinent – US$900,000 that had been allocated to China on its becoming a member of the WHO, and which China, having already solved its own smallpox problems, had just refused.[29] But a smallpox crisis in north-east India soon followed, as the outbreaks had not been sufficiently suppressed in the autumn season. In Bihar the number of reported smallpox cases doubled in February, and in March there were more than 12,000. The situation was not helped by an airline strike, a rail strike, a drought in southern Bihar and floods in northern Bihar – both drought and floods generating refugees. In May the number of cases of smallpox reported was over 35,000, and it

was only with difficulty that the Director of Health Services was persuaded to give the containment strategy one last month before switch-
ing to mass vaccination.[30]

In Uttar Pradesh, there had been little change during January and
February, but the reported numbers climbed gradually in March and
April, peaking at about 8,000 in May. To encourage villagers to report
cases, one district magistrate led a parade of 5,000 Congress Youth
volunteers, with an elephant covered in appropriate brightly coloured
slogans.[31] From both Bihar and Uttar Pradesh, smallpox spread to
West Bengal, Orissa, Madhya Pradesh and Maharashtra.

A particular source of the spread from Bihar was Jamshedpur, an
industrial complex containing heavy industries of the Tata group, and
one of India's most important steel-making areas, whose prosperity
attracted both seasonal workers and beggars from Madhya Pradesh and
Orissa. Many of the population lived in company towns or railway sid-
ings, the health service was poorly organised and fragmented, and no
one was responsible for reporting cases of smallpox among the migrant
workers, who anyway tended to travel back to their homes when they
were ill.

The member of the WHO Indian team sent by Nicole Grasset to
cope with this problem was Lawrence Brilliant. He was an unusual
member, a leftover from the hippie trail, who had come from California shortly after qualifying as a doctor – one of a bus-load of enthusiasts bringing relief aid to East Pakistan. He had stayed on for a life of
peaceful meditation in a Himalayan monastery, but this all changed
when, encouraged by his guru, he committed himself instead to the
strenuous life of a field worker in the smallpox campaign.

Brilliant and Nicole Grasset both realised that the key to the situation in Jamshedpur lay with a few powerful men. Brilliant approached
the managing director of the Tata Iron and Steel Corporation, roused
his interest, and persuaded him to get in touch with J. R. D. Tata
himself in Delhi to tell him what was going on. Next Nicole Grasset
and Brilliant went to see Tata, who was shocked into promising nearly
US$1 million and assembling doctors and helpers and transport to

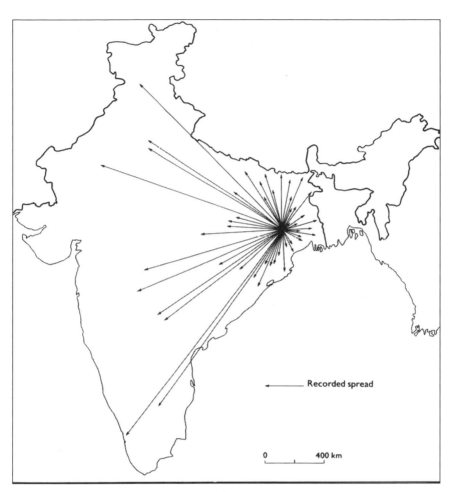

33. Map of India showing the spread of smallpox from Jamshedpur in 1974

search out smallpox in Jamshedpur and adjacent districts in southern
Bihar. With this support, and the help of the government of India, the
state of Bihar, OXFAM and other charities, the outbreak was tackled by a vigorous containment programme involving the blocking of
major roads and bridges, the diversion of trains to special inspection
areas, checkpoints at bus stations and house-to-house vaccination. The
whole city of 600,000 was quarantined, and Brilliant has described
how all surrounding villages were vaccinated, however remote and
with whatever hazards of dacoits or marauding elephants.[32] A tribal

leader protesting that 'only God can decide who gets sickness and who does not' echoed the fatalism of Edmund Massey preaching 150 years earlier against inoculation, but in a few months the outbreak was brought under control. More than 2,000 cases of smallpox in eleven states are reckoned to have arisen from it.

When the Bihar epidemic was at its height, away to the west in the Rajasthan desert India exploded its first nuclear device. The foreign press made much of this embarrassing mixture of technological advance and medical chaos.

After the frightening number of cases in May, a dramatic fall in June with the onset of the monsoon rains was especially welcome; but the crucial question was whether intense efforts could prevent, or at least minimise, a repeat rise in the number of cases in the new year. The central authorities therefore agreed to double the number of epidemiologists in the field (some of these came from the CDC in Atlanta), and to create six new centrally controlled surveillance teams and 300 extra containment teams. They bought 375 extra Jeeps; they kept a register of all people living near infected houses; they arranged for more rapid vaccination over more extensive areas, and they offered greater rewards for reporting cases. The cost of all this was met partly by a generous gift from the Swedish International Development Authority, and partly by increased central government grants.

In the course of 1974 nearly 127,000 cases of smallpox had been reported in Bihar, about two thirds of the total for India; but in December there were only about 500. There was a setback, though, at the end of the month when refugees from famine following autumnal floods in northern Bangladesh brought smallpox into eastern India. Measures to tackle this even included night searches among the 48,000 street dwellers in Kolkata. There were outbreaks, too, at pilgrimage sites of the Jain religion – particularly severe in the town of Puri, in Bihar, where the founder of the Jains, Mahavira, had died two and half millennia earlier. These outbreaks were worrying since Jains, concerned for the pain suffered by cows in the preparation of vaccine, tend to be reluctant to accept vaccination; fortunately their religious leader was persuaded to recommend it, the town was quarantined by military police, strict rules of containment were enforced, and by the end of February 1975 the outbreak was over.

After this massive effort, the situation looked hopeful. Though there had been just over 1,000 cases reported in January, there were only 212 in February, and fewer than ninety in March and in April. May saw the last case in India – Saiban Bibi, a homeless thirty-year-old Bangladeshi beggar living on a railway platform in Karimganj, a small town in Assam, who had caught smallpox from a patient in nearby Sylhet. For four days she had lain shivering with fever, either on the platform or in the third-class waiting room, moving only to get tea from a nearby stall. Feeling worse on the morning of the sixth day, she had been advised by passengers to go to the Karimganj Civil Hospital. Smallpox was diagnosed and in a rapid operation 27,000 people were vaccinated in Karimganj alone. Happily, Saiban Bibi recovered, and despite the large number of people she must have come into contact with there were no further cases.[33]

The last case in India was not the last case in the Indian subcontinent. Smallpox had by then been eradicated from Afghanistan, Pakistan, Nepal, Bhutan and Sikkim, but it still flourished in Bangladesh. Ironically, back in August 1970, despite being one of the poorest and most densely populated countries in the world, Bangladesh (then East Pakistan) had succeeded in eradicating smallpox, at the end of a campaign in which mass vaccination had been followed by eight months of effective surveillance-containment.[34] For six months, including the cool dry season from November to March, there were no cases. March 1971 saw the start of the civil war that led to the independence of Bangladesh. Of the population of 68 million, about 10 million fled to India; even more left their homes for other parts of the country. There was a great deal of violence and famine, and health services were disrupted, but even by the end of December – sixteen months after eradication had been achieved – there was still no smallpox.

The 10 million refugees who had fled to India presented a massive problem to the Indian authorities. Many were screened for smallpox, housed in special camps and vaccinated. But in West Bengal, officials did not permit visits to the camps by national or WHO health staff; and medical care in the largest camp – Salt Lake Camp near Calcutta – was left to a voluntary agency which largely ignored vaccination.[35] Cases of smallpox seem to have appeared in November and been misdiagnosed as chickenpox; William Foege, back home in Atlanta watching a television film of the camp and its hospital, was horrified to recognise what was clearly smallpox.[36] After a chain of telephone calls and telexes, and a denial by the West Bengal Director of Health Services, the diagnosis was confirmed by an officer of the Indian government, and isolation and vaccination were started.

This, though, was locking the stable door after the horse had bolted. On 16 December 1971, Bangladesh had become independent, and by mid-January about 50,000 refugees had already left to go home, many of them taking smallpox with them. They were travelling in crowded conditions in the cold, dry season when transmission is easiest, and they were returning to a country whose services and facilities had just been disrupted by a vicious civil war. An epidemic was inevitable and, though surveillance teams were swiftly set up and large numbers of vaccinators hired, it is estimated that there were about 90,000 cases in Bangladesh in 1972.[37] Famines, floods, and slum clearance that uprooted and scattered between fifty and a hundred thousand people, aggravated the situation. It was not until November 1975, after a heroic effort involving over 24,000 staff (including more than 200 international staff from twenty-eight countries),[38] that the last case of smallpox in the Indian subcontinent was detected – a three-year-old girl, Rahima Banu, from Bhola Island in the mouth of the Ganges, who made a good recovery.[39] She was also the last case of naturally occurring variola major anywhere in the world.

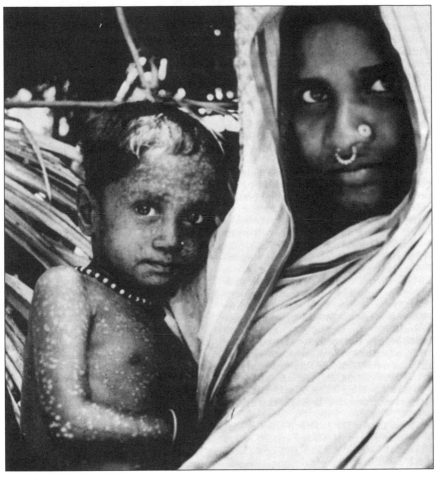

34. Rahima Banu, the last case of naturally occurring variola major in the world

By December 1975, smallpox existed only as variola minor, and only in Ethiopia. Just when (and why) variola major disappeared from Ethiopia and variola minor appeared, is not known. There is ample evidence of the presence of variola major in the accounts of foreign travellers and in Ethiopian records from the seventeenth century to the 1920s.[40] Variolation was widely practised in the early eighteenth century, and probably long before; vaccination against smallpox was used in Eritrea, under Italian influence, in 1889, and in Addis Ababa, under French influence, nine years later.

Until 1970, Ethiopia had refused to become involved with the intensified smallpox eradication campaign, partly because of the relative
mildness of the current disease, and partly because of the intransigence
of health workers involved with the eradication of malaria, who neither
welcomed a rival scheme nor were prepared to collaborate in running
it. The situation changed when Emperor Haile Selassie was persuaded
by Kurt Weithaler, the Austrian director of his hospital, of the value
of getting rid of smallpox, and early in 1971 plans were made both for
surveillance and for mass vaccination. Weithaler was put in charge of
the campaign, helped by the WHO epidemiologist Ciro de Quadras,
who had worked in Brazil, and the Ethiopian Ato Tamiru Debeya.

Ethiopia's population was extraordinarily widely dispersed, and
spoke so many languages that staff of the eradication programme often
had to communicate through a succession of interpreters. Over half
of its 25.5 million people lived more than a day's walk from any road,
most of them in the central and western highlands, whose deep ravines
were often impassable during the long rainy season; less than 10 per
cent lived in communities of more than 500 people. For all these reasons, it was hoped that if smallpox could be eliminated from the more
densely populated areas, transmission chains would not be maintained
indefinitely, and the disease would simply die out. In fact, eradication
came only after five eventful years.[41]

From the start there were far more cases of smallpox than had been
expected – so many, in parts of the south-west, that village-by-village
containment was reckoned 'futile' and all that could be done was to try
'to define the outer limits of the spreading epidemic' and to concentrate on vaccinating the people there, 'much as one would fight a forest
fire'.[42] In March the disease crossed the southern border when an infected Ethiopian cattle herder took his herd into smallpox-free Kenya and
caused an outbreak of forty-six cases. By the end of 1971, the number
of cases of smallpox reported in Ethiopia was over 26,000 – 10,000
more than were reported in the whole of India (with twenty times the
population) during the previous year. Fortunately, because the cause
was variola minor, the proportion dying was only about 2 per cent.

By the end of 1972 about 6 million people had been vaccinated, particularly in the south-west of the country where more than half the population were now protected, and matters were improving. But there was resistance from two large ethnic groups; and though it was sometimes possible to persuade people to be vaccinated by the offer of other medical services, members of one eradication team were even bitten when they tried to vaccinate.[43] It became apparent, too, that variolation was still common, and that as many as 12 per cent of the cases of smallpox might be attributed to it. Luckily, not all the ethnic groups that practised variolation rejected vaccination.

In the autumn of the following year, a severe drought in the eastern and north-eastern parts of the country led to a famine that killed about 200,000 and caused massive movements of people, some carrying smallpox from endemic areas in the north-east to smallpox-free regions elsewhere in Ethiopia, and across the border into Somalia and Djibouti (then the French Territory of the Afars and Issas). Despite this setback, the total number of cases reported in Ethiopia in 1973 was fewer than 5,500. Early in 1974, substantial short-term help was provided by the French, the Sudanese and the Kenyans, and in June the USA Public Health Service provided funds for the use of two small helicopters.

1974 also saw the start of a revolution which removed Haile Selassie and the whole feudal structure of land tenure. It was triggered, in part, by the failure of the authorities to cope with famine; and it was accompanied by eruptions of chronic smouldering violence in Eritrea in the north, among the Amharic people in the central highlands, and between Ethiopians and Somalis in the Ogaden desert in the east. Yet by the end of the year the annual number of smallpox cases had fallen by nearly a fifth, many regions were free of the disease, and the only remaining large area of endemic smallpox was in the mountainous central zone. The helicopters were captured and shot at and temporarily lost – one of them was destroyed by a grenade – but they made it possible to reach people in the almost inaccessible highlands, even if their limited loads meant that fuel dumps had to be arranged beforehand, and teams sometimes needed to walk or ride mules for up to ten days from the landing point to their destinations.

1975 began ominously, with many infected areas cut off by civil war, but the eradication of smallpox first in India and then in Bangladesh freed resources that could be redirected to the campaign in Ethiopia – a campaign that looked as though it would be the last in the global eradication programme. Until 1974 there had been fewer than ninety eradication staff and a handful of advisers from the WHO; by 1976 the total number of staff had grown to over 1,000, including twenty-five from the WHO.

The new government proclaimed that the eradication programme had the highest priority, more helicopters were made available, and elaborate measures were taken to ensure supplies of petrol, money and vaccine. Two areas caused particular concern. The first was the central highlands including the provinces of Gonder and Gojam and the vast Blue Nile Gorge. (Gojam is the province in which, 200 years earlier, Bruce had described the custom of containing smallpox by thrusting those with the disease into the flames of their burning houses. That was no longer the custom, but Gojam was again a violent area where two vaccinators had been killed and searching for cases was risky.) The other area was the Ogaden desert, where smallpox persisted among the nomads and the settled agricultural population with whom they were in contact.

Despite the difficulties, the last cases in the central highlands were discovered in early July. Later that month an outbreak was detected in the nomad village of Dimo in the southern part of the Ogaden – it had apparently been going on for some weeks but fight-ing between Ethiopian and guerrilla forces had previously made the area inaccessible. There were sixteen cases, the last being a three-year-old girl, Amina Salat. She had been variolated, and was vaccinated while she was incubating the disease, but too late to gain immunity. She recovered, and for some weeks was thought to have been the last case of smallpox in the world. But late in September 1976, when preparations had already been made for a celebratory press conference, there were reports of cases of smallpox in Mogadishu. The disease had spread to Somalia.[44]

———

Though this was a great disappointment, it did not at first seem too alarming. By the end of September there were only five known cases in Mogadishu. All five were supposed to have been living in one or other of two sites in the southern Ogaden within Ethiopia until eight to ten days before they became ill, when they began their journeys to Somalia. This pointed strongly to the persistence of smallpox in the Ethiopian Ogaden, and as both the sites named were in areas in which guerrilla activity had interfered with surveillance, the story seemed plausible. Somalia had been free of endemic smallpox since 1963, and because 85 per cent of the Somali population were supposed to have been vaccinated the risk that the imported cases of smallpox would give rise to an epidemic there seemed slight.

There had, though, already been some doubts at the WHO headquarters about the reliability of Somali reports. It seemed too good to be true, for example, that of thirty-eight imported cases of smallpox in the three and a half years preceding February 1976, only one had caused secondary cases.[45] Much later it became clear that there had been a failure to report outbreaks of smallpox several months before the five cases in September, the Somali staff hoping to contain them without embarrassing publicity. By the end of October twenty cases had been reported in Mogadishu, but WHO epidemiologists were not allowed to question patients in hospital or to assist local health staff in containment procedures. By the end of the year, the total number of cases admitted to hospital was said to be only thirty-four. An unofficial hospital register would later reveal there had been over 500.

Matters improved when a new manager of the Somali Smallpox Eradication Programme, Abdullahi Deria, was appointed. Working with Ehsan Shafa of the WHO Smallpox Eradication Unit, he started an intensive search in March 1977, offering a reward of 200 Somali shillings (US $32) for each case of smallpox reported. And the reward may have done its job, for a few weeks later two cases of smallpox were reported from southern Somalia not far from the Ethiopian border. In April there were more than 150 cases in over forty separate outbreaks spread over most of the administrative districts in the southern half of the country. In May, with over 600 reported cases, there were serious worries that Kenya and Ethiopia might be reinfected and, more fright-eningly, that if the epidemic was not stopped before the autumn the disease could be carried by pilgrims to Mecca.

The government declared the situation a disaster, and appealed for help to the United Nations Disaster Relief Organization. By early June contributions worth more than US $400,000 in cash or kind began to arrive – from donors including Canada, the Netherlands, Norway, Sweden, the UK, and the Red Cross and Red Crescent. Air-lifts brought Land Rovers, camping equipment, radio-telephones. By this time Henderson had left the WHO to become Dean of the School of Public Health at Johns Hopkins University, in Baltimore, and Isao Arita had taken over as head of the Smallpox Eradication Unit – the unit that, at its birth, had consisted solely of him and a secretary. He had already made a number of visits to Somalia, and in April had increased the number of WHO epidemiologists there to fifteen. By July there would be twenty. The chief WHO adviser in Somalia was Zdenek Jezek, a Czech epidemiologist who had worked with Grasset, Foege and Brilliant in India, having earlier spent five years setting up the first national health programme in outer Mongolia. He arrived in Mogadishu on 10 May and eleven days later the Somali government agreed on a detailed emergency programme.

The problems were formidable. There were not enough team lead-ers, or surveillance agents or vaccinators. Special techniques involving criss-cross paths had to be developed to ensure that a few two-person surveillance teams could reliably search areas for nomads, who were difficult to find in the high scrub. Often only 10 per cent of the nomads had been vaccinated, and the offer of vaccination, though usually wel-comed, sometimes made them flee. An unexpected complication was that the nomads often had self-inflicted wounds on their upper arms, which made checking for vaccination scars unreliable.[46] Guerrilla war-fare between Somalia and Ethiopia was increasing and in some areas radio communication was forbidden by the Somali authorities. The rains between March and June that year were particularly heavy, wash-ing away roads. In the field it was necessary to provide tents and to build thorn barricades around campsites at night to keep out hyenas. And the number of cases of smallpox was increasing every week.

Nomads added to the isolation problem, too, for instead of being kept in their own homes with watchguards, they had to be removed to isolation camps. Those with variola minor, though, were not incapaci-tated, and because the camps tended to be out of the way and not com-fortable, patients often left while they were still infectious. One solution was to pay them five Somali shillings a day while they remained, and to give them new clothes when they had recovered, burning the old ones. Another was to construct circular barriers of thorn bushes, this time to enclose a latrine and a lean-to shelter in which patients could be isolated without leaving their community.

The epidemic peaked in June, when over 1,300 cases were reported; by then the number of staff involved in the eradication programme had increased from a little over a hundred in March to over 3,000 – the result of a heroic training programme. And there were hopeful signs: individual outbreaks were tending to involve fewer cases and to be over more quickly. In southern Somalia, the proportion of the population who had been vaccinated reached 70 per cent by August and 90 per cent in September. The number of cases fell sharply.

The very last case was Ali Maow Maalin, a 23-year-old hospital cook in the banana port of Merca, on the Indian Ocean about fifty miles south-west of Mogadishu. Despite his job, and although he had at one time worked as a vaccinator, he had never himself been vacci-nated satisfactorily. In October a health officer drove a mother and her two sick children into the town seeking the smallpox isolation camp. Not knowing where it was he drove to the hospital, and Maalin kindly climbed into the Land Rover to direct them. The journey was short and he was in the car for only a few minutes. Nine days later he felt

35. Ali Maow Maalin, the last case of naturally occurring smallpox in the world

ill and feverish and went home, where he was visited by friends and neighbours. Three days after that he was admitted to hospital, where he was thought to have malaria, until a rash appeared and the diagnosis was changed to chickenpox. It was only after he had been discharged and sent home that smallpox was diagnosed, by a visiting nurse from the hospital, and confirmed by a WHO epidemiologist. The diagnosis caused a storm of activity: nearly 55,000 vaccinations were done in the next fortnight, over ninety contacts were followed up, the town of Merca was searched weekly for six weeks, and the surrounding region monthly for five months. No more cases were discovered. Maalin made a complete recovery. One of the two children, a little girl, died.

Though Maalin was the last case of smallpox caught in the natural way — and so the end of a chain of infection stretching back at least 3,000 years — he was not, sadly, the last case of smallpox. Nor was the little girl the last fatality. Less than ten months after Maalin's recovery a forty-year-old photographer in the medical school in Birmingham, England, was found to have smallpox caused by variola major. She worked on the floor above the department of microbiology — a department which specialised in studying smallpox — and the infection almost certainly came through a service duct connecting the two floors. Four months earlier WHO inspectors had criticised procedures in the microbiology department. Remembering this, and shattered by the tragedy, the distinguished head of the department committed suicide. The photographer died later; during her illness her father died of a heart attack, and her mother became ill with smallpox but ultimately made a full recovery.[47]

16

'And out of good still to find means of evil'

The words are Satan's, in Milton's *Paradise Lost*.

On 9 December 1979, the global commission that had been set up to certify the eradication of smallpox agreed that the job had been done. For a quarter of a century now, there have been no cases of smallpox anywhere. Yet we read in newspapers that the governments of the USA, the UK and many other countries have arranged to buy vast amounts of smallpox vaccine, and that shortly before the Iraq war emergency personnel in America, in Britain and in Israel were vaccinated against smallpox. The highly respectable *New England Journal of Medicine* devoted nine pages of its April 2002 issue, and more than thirty pages of its January 2003 issue, to various aspects of smallpox and vaccination. And the only official remaining stocks of variola virus (in Atlanta in the USA, and in Koltsovo in Russia) continue to be preserved.

For though eradication was recognised by all countries as a stunning success, a few countries, and in particular the USSR, saw it as an opportunity. In a world in which most people have either never been vaccinated, or have not been vaccinated for many years, we are likely to be as vulnerable to smallpox as the Native American Indians were in the seventeenth century – which makes smallpox a horribly tempting weapon for biological warfare. And this would not be a matter of leaving infected blankets.

Biological warfare has a long history, going back far further than

36. *The official parchment certifying the global eradication of smallpox,*
9th December 1979

seventeenth-century America. Whether by poisoning wells with dead animals or catapulting plague-ridden bodies over the walls of a besieged city, we have ever been ready to exploit all the horrors that nature offers. The resources of biology played their part in the revolution in all types of warfare in the twentieth century. There were rumours that the Germans were trying to spread plague in Russia in 1915. Ten years later Winston Churchill wrote:

> A study of Disease — of Pestilences methodically prepared and deliberately launched upon man and beast — is certainly being pursued in the laboratories of more than one great country. Blight to destroy crops, Anthrax to slay horses and cattle, Plague to poison not armies only but whole districts — such are the lines along which military science is remorselessly advancing.[1]

Once naturally occurring smallpox had been eradicated, and vaccination against it ended, it could be added to the list.

The prohibition of 'bacteriological methods of warfare' in the Geneva Protocol of 1925 alerted some to the idea that such methods must be worth pursuing. The Russian biological warfare programme began in 1928 with work on typhus.[2] In Japan, Shiro Ishii set up a germ warfare research centre (the Pingfan Institute) in 1935, which experimented with various forms of germ warfare, including smallpox, over the next ten years[3] — research which the Russians learnt about in 1945 and decided to follow in a new biological complex at Sverdlovsk.[4] The history of the Second World War has plenty of rumours of biowarfare atrocities — of Russians spreading cholera and anthrax, and letting tularemia loose to break the siege of Stalingrad;[5] of Japanese using prisoners as human guinea-pigs, raining plague-fleas on China and using aerosolised variola virus on Chinese prisoners of war in Manchuria.[6] The Stockholm International Peace Research Institute confirmed the Japanese atrocities: 'A conservative estimate is that the Japanese biological warfare programme killed at least 10,000 people in its laboratory experiments and perhaps as many as several hundred thousand others in military field operations.'[7] The experiments were carried out on political opponents, the mentally handicapped, prisoners — anyone considered to be expendable. Russians and Poles are both said to have found germs useful in guerrilla warfare.[8] It is not surprising to learn that, during the Second World War, the British thought about the possibility of using smallpox, and started investigating (but never used) anthrax. The United States, while saying that biological warfare was 'dirty business',[9] decided to be prepared, and Fort Det-rick, in Maryland, became the centre for American biological weapons research.

In 1969, President Nixon declared: 'The United States shall renounce the use of lethal biological agents and weapons, and all other methods of biological warfare.' Three years later Russia agreed to do the same. The United States did still have chemical and some biologi-cal weapons, though there is no evidence that they had ever worked on variola virus as a weapon; they abandoned their factories and publicly destroyed at least some of their stocks,[10] while continuing research on vaccines and anti-viral drugs.

But the Cold War was a time of appalling cynicism and deception, and Russia kept the smallpox factory at Zagorsk fully active. Here eggs were injected with variola virus, sealed with paraffin, then incubated until the virus took over its host, and the liquid in the egg was ready to be stabilised and stored. The process was, by chance, improved in 1959 when a traveller from India, vaccinated only in childhood and suffering symptoms of a particularly vicious strain of smallpox, infected forty-six people in Moscow – prompting the Russians to collect and treasure the strain for possible future use: 'Within a few years, India's unwit-ting gift became our principal battle strain of smallpox.'[11] The traveller had arrived just a year after the World Health Assembly at which the Soviet delegate Victor Zhdanov had proposed a five-year plan to eradi-cate smallpox. Twelve years later, on Rebirth Island in the Aral Sea, the Russians experimented with releasing variola virus as an aerosol.* By the 1970s, a stockpile of twenty tons of the virus was being kept at Zagorsk – matched by twenty tons of plague bacillus kept at Kirov.[12] The idea of measuring plague or smallpox in *tons* is terrifying.

---

*In June 2002 the Monterey Institute of International Studies in California published a translation of a secret 1971 Soviet report of an outbreak of smallpox in the city of Aralsk, on the north coast of the Aral Sea. It had started with a young female fisheries officer on a

In 1973, breaking away from the stranglehold of Lysenko's maverick genetic views,[†] the Soviet Union set up Biopreparat, ostensibly a state-owned pharmaceutical complex with research centres all over Russia, but in fact an agency for developing and genetically engineering biological weapons – an agency that at its peak employed about 30,000 people, while 30,000 others worked in biological warfare laboratories scattered through the country.[13] An extra – and, as it turned out, unfortunate – push was given by the United States, who deliberately misled the Russians into believing that biological weapons were still being developed in America. Intended to lead the Russians into useless research, the policy backfired.[14]

In this web of deceit Zhdanov, of all people, was appointed chairman of Biopreparat. After two years he was made to resign, though whether this was because he was out of sympathy with the work or because he was incompetent as a manager is not clear. While variola virus was being stockpiled at Zagorsk, Zhdanov became director of the Ivanovsky Institute of Virology in Moscow, where he spent the rest of his life in virtuous research on hepatitis, influenza and AIDS; his wife claims that he spoke out against biological weapons.[15]

Soviet double-dealing was perhaps most marked in 1980, the year when the World Health Assembly recommended that vaccination of civilians should end – and the year when the Kremlin launched a five-year plan for the intensive study of certain viruses, including variola, with the aim of producing more lethal strains. Particularly chilling was the idea of developing a strain resistant to ionising radiation.[16] There

---

research vessel that had been about nine miles from Rebirth Island, and it had spread to her schoolboy brother and to eight others, of whom three unvaccinated individuals – a teacher who had visited the brother, and two infants under a year old – died. The official Russian explanation was that the infection must have originated in Afghanistan and that the fisheries officer must have caught the virus when the ship docked at ports in Uzbekhistan or Kazakhstan. But the officer had not disembarked, and an alternative explanation is that the source of the infection was the smallpox weapon tests on the island.

†Trofim Lysenko (1898–1976) was Stalin's 'dictator in biological sciences', who condemned conventional genetics as 'bourgeois science', and dominated Russian biology from the mid-1930s until 1965.

is a depth of evil and hysteria in the thought that in the aftermath of a nuclear war smallpox should still be there to attack survivors. Russians were no longer protected by vaccination, so if it spread there would be no winners.

There were plenty of rumours in the West about Soviet weaponry. In October 1979 the British magazine *Now!* was the first to announce 'Russia's secret germ warfare disaster':

> *Hundreds of people are reported to have died, and thousands to have suffered serious injury as a result of an accident which took place this summer in a factory involved in the production of bacteriological weapons in the Siberian city of Novosibirsk.*[17]

The accident was in fact not in Novosibirsk, but in the Urals, in Sverdlovsk (now Yekatarinburg), and the germ was anthrax – a 'natural outbreak of anthrax among domestic animals', the official Soviet news agency TASS said. Although the number of people killed is uncertain there is no doubt that the accident was at a plant designed to produce anthrax as a weapon. The Communist Party chairman of Sverdlovsk at the time, responsible for cleaning up and covering up the incident, was Boris Yeltsin; in 1992 he admitted that there had been at least sixty-six deaths.[18]

Quite apart from the risks of weaponisation, it was obviously desirable to reduce to a minimum the number of laboratories holding stocks of variola virus. Even before the completion of the eradication programme, the WHO had succeeded in reducing that number to eighteen, and by the end of 1981 it was down to five – in the USA, the Soviet Union, Britain, China and South Africa. The tragic disaster in Birmingham in 1978 made those running smallpox laboratories less reluctant to destroy inessential stocks of variola. One cogent reason for retaining the virus was to clone fragments of the DNA of different strains so that libraries of fragments could be stored (without any risk) and be available to identify the cause of any future outbreaks. After the virologist Keith Dumbell had treated samples of all the variola viruses in the British collection at Porton Down (near Salisbury) in this way, he took the remaining stocks – in 200 vials packed in dry-ice – to Atlanta for storage in the CDC. In 1983, the Chinese reported that they had destroyed their stocks, and the South Africans agreed to destroy theirs after the library of cloned variola DNA fragments from the Porton collection had been made available to them.

So then there were (officially) two – the USA collection at the CDC in Atlanta, and the Russian collection in the Ivanovsky Institute of Virology in Moscow (later transferred to the Vector Institute at Koltsovo, near Novosibirsk). Whether there were also unofficial secret stores of smallpox elsewhere remained an open question and a major worry.

In 1987, ten years after the last naturally occurring case of smallpox, Kanatjan Alibekov, soon to be chief scientist and deputy chief of Biopreparat, had been asked to turn from tularemia and anthrax and, under Mikhail Gorbachev's five-year plan, to supervise the creation of a new smallpox weapon. Within three years, the weapon, a small 'bomblet' which released a suspension of variola virus as an aerosol, was being tested inside a specially designed chamber at the Vector Institute. Alibekov calculated that the Koltsovo production line was capable of manufacturing 80–100 tons of variola virus a year. Large numbers of the bomblets could be delivered either in cluster-bombs dropped by long-range aircraft or in intercontinental ballistic missiles, and he hoped they could be used with the genetically altered strains that were being developed.

We know all this through disillusioned scientists who decided to defect and tell the West what was going on. In 1989 the Soviet government was appalled to learn of the defection of Vladimir Pasechnik, who had been working on a project to modify cruise missiles so that they could deliver biological agents[19] – and the British and American governments were appalled to hear his news about the Biopreparat complex and its stores of variola virus and genetically engineered, antibiotic-resistant plague bacilli.[20] Protests were met by a denial from Gorbachev; British and American leaders decided to keep the whole affair quiet for the time being, possibly for the sake of good relations with Gorbachev during difficult negotiations about arms controls. Then in 1992, after the collapse of the Soviet Union, after Yeltsin had banned offensive biological research, and after the Biopreparat research centres had had their funds severely cut, Alibekov himself defected to the United States. What he had to say was so shocking that, he claims, it was a struggle to persuade the Americans how far the science of germ warfare had come.[21] The results of his year-long debriefing were kept secret, but in 1998 he spoke out on American television and to the press. In 1999, Americanised as Ken Alibek, he published a chilling account of his life.

In spite of the changes that followed Yeltsin's decree – which includ-ed the ending of research on bio-weapons, the destruction of assem-bly lines at various sites, and the conversion of bio-weapon facilities into pharmaceutical and pesticide plants – much of the former Soviet Union's offensive weapons programme was, according to Alibek, still viable. In 1993, he said, Soviet scientists had succeeded in inserting for-eign genetic material into vaccinia as a prelude to doing the same trick with the variola virus. The insertion of foreign genetic material into vaccinia had been achieved a decade earlier by workers in the USA seeking to make new vaccines against a variety of diseases, but the aims of the Russian project were quite different. As well as making strains of variola resistant to existing vaccines, and strains containing gene coding for various protein toxins, they planned to insert into the variola virus genetic material from the Ebola virus or from the incapacitating Venezuelan equine encephalitis virus, so creating an agent that would cause two lethal diseases at the same time.[22] (Whether creating such chi-meras is in fact feasible is uncertain, though there is a claim that scien-tists at Koltsovo succeeded in making 'a chimera of Venezuelan equine encephalitis and mousepox viruses that was stable and, when used to infect mice, appeared to cause the symptoms of both diseases'.[23])

Even if the stockpiles of bio-weapons no longer existed, Alibek

235

warned, there was still plenty to worry about in the fate of stocks of dangerous materials, and microfilms of detailed instructions. Vials of biological agents, small and easily smuggled, were 'rumoured to be circulating freely in the Russian criminal underworld'.[24] For smallpox, the worry was not so much the official stores of variola virus at Koltsovo (the Vector Institute there was no longer a secret laboratory), as the still closely guarded Virology Institute at Zagorsk.[25] And there was no way of knowing whether samples of the virus that had been kept at other institutes during the many years of active bio-weapons research had all been destroyed.

Alibek was also concerned about the current activities of former Biopreparat scientists outside Russia. He reckoned that at least twenty-five of them were in the USA, but many more were in Europe or Asia – several had gone to Iraq or North Korea, and at least five to Iran. A 1993 KGB report had added to the rumours of North Korean research into smallpox for biowarfare.[26]

In 1995 the CIA and the US Defense Intelligence Agency were worrying that there might be as many as eight countries with undeclared stocks of smallpox virus. Yet it was not until 2001 that instructions for weaponising a number of agents, including smallpox, were removed from the internet.[27] In early November 2002, during the build up to the war against Iraq, the *Washington Post* reported comments from anonymous officials, said to be based mainly on a briefing from the CIA's Weapons Intelligence, Nonproliferation and Arms Control Center (WINPAC), that implied that there was good evidence that four countries – Iraq, North Korea, Russia and France – had secret stores of smallpox.[28] The WINPAC report also claimed that a former Soviet scientist had told US officials that smallpox technology had been transferred from the USSR to Iraq in the early 1990s. In October 2003 Donald Henderson, now increasingly suspicious, told a conference in Geneva that Iraq, Syria or Iran might have kept stocks of smallpox after the widespread outbreak in 1970–72 that had started with pilgrims from Afghanistan.[29] And on 3 November 2003, the CIA issued a report – ominously entitled 'The

Darker Bioweapons Future' – warning that the increased sophistica-
tion of genetic engineering could lead to more dangerous biological
weapons. They mentioned, as an example of the kind of technique
that might be used, the increased virulence inadvertently produced in a
mousepox virus by researchers in Australia who had inserted into it an
immunoregulator gene.

But what about the official stores of variola virus at Atlanta and Koltsovo? Arguments about their future have gone on for twenty years.[30] In
1990 the US Secretary for Health had optimistically announced that,
after completing the sequencing of the genome of the virus responsible for smallpox, the United States would destroy all remaining
stocks; and he invited the Soviet Union to do the same so that they
could 'jointly announce the final elimination of the last traces of this
lethal virus'.[31] The United States proposed doing their part in 1996;
after much discussion they postponed it for three years and then, with
increasing knowledge of what was going on in Russia, they abandoned
the idea because of:

> our concern that we cannot be entirely certain that after we destroy
> the declared stocks in Atlanta and Koltsovo, we will eliminate all the
> smallpox in existence. While we fervently hope smallpox would never
> be used as a weapon, we have a responsibility to develop the drug and
> vaccine tools to deal with any future contingency – a research and devel-
> opment process that would necessarily require smallpox virus.[32]

The case for destruction is further weakened by public knowledge
of the complete DNA sequence of various strains of variola virus.
Though current techniques are not good enough to synthesise the virus
from scratch (or by modifying another orthopox virus), it would be
surprising if such a synthesis did not eventually become possible, so
destruction of stocks would not be a permanent safeguard. (How soon
'eventually' is, is not clear. In work at Stony Brook supported by the

USA Defense Advanced Research Projects Agency, the polio virus was recently synthesised from its components using knowledge of its RNA as a blueprint; but that process took two years and the variola virus is much bigger and more complex than the polio virus.[33] On the other hand, more recently, it has been reported that Craig Venter and his colleagues took only two weeks to put together pieces of synthetic DNA bought from a biotechnology company to make a known virus that attacks bacteria and that has a genome about two-thirds the size of that of the polio virus.[34])

The strongest positive argument for retention is that live variola virus is necessary for the development and testing of better vaccines and of anti-viral drugs effective against smallpox. If there ever were an outbreak of smallpox in an unvaccinated world, such drugs would be helpful in the early stages of the outbreak — and also later for people who could not safely be vaccinated, either because they had eczema or because their immune systems were impaired. Though during the millennia in which smallpox flourished no cure was discovered, in the last few years there has been some progress in finding one. John Huggins and his colleagues from Fort Detrick, using facilities at the CDC, reported that a drug called *cidofovir*,[35] which is routinely used to treat a disease of the retina caused by a type of herpes virus (and is known to be active against the vaccinia virus), is also highly active against the variola virus. Cidofovir has to be injected into a vein and it has unpleasant side effects including damage to the kidneys; but Karl Hostetler and his colleagues at San Diego and at Birmingham, Alabama, have synthesised a modified form of cidofovir that can be given by mouth, that is not toxic to kidneys, and that is more than a hundred times as effective as the unmodified drug when tested on variola virus growing on cultured cells.[36] It also prevented death in a mouse given a lethal dose of cowpox; whether it would be effective in treating a patient with smallpox it is impossible to say.

A quite different strategy for developing a therapy for smallpox may arise from recent experiments by Ariella Rosengard and her colleagues in Philadelphia.[37] Believing that the restriction of smallpox to humans might reflect the possession by the variola virus of proteins particularly effective at interfering with the human immune response, they compared proteins from variola and from vaccinia that interfere with the human 'complement system' – the battery of proteins that destroys complexes of foreign proteins with their antibodies. They found that the variola virus does indeed contain a protein that is 'particularly adept at overcoming human immunity', and they point out that disabling that protein – they call it the 'smallpox inhibitor of complement enzymes', and give it the acronym SPICE – could be an effective approach to the treatment of smallpox.

A remarkable feature of the work on SPICE is that it did not involve working with the variola virus: instead the Philadelphia workers 'molecularly engineered' the protein, knowing the structure of the DNA that encodes it. But other approaches to finding anti-smallpox drugs or to improving smallpox vaccines do require the use of live virus, and it is not surprising that the WHO secretariat's Advisory Committee on Variola Virus Research, meeting near the end of 2002, felt 'further research was still needed before consensus could be reached on a date for the destruction of the remaining stocks of the virus'.[38] Research using live virus should 'continue to be carefully monitored and reviewed under the auspices of WHO'.

It is easy to postpone decisions about the future of the official stocks of variola virus; it is less easy to postpone decisions about the best way to deal with the threat of future infection, either by variola or other pox viruses. The difficulty is in balancing risks posed by the disease against risks posed by measures that might be taken to avoid it. Of course this is not a new situation. With hindsight we feel that Leopold Mozart probably made the wrong choice when he refused to have Wolfgang variolated in Paris; forty years earlier, James Jurin had been more hard-headed about the relative risks, but it is unlikely that Leopold had read the *Philosophical Transactions of the Royal Society*. The choice became much easier after Jenner, because the risks of vaccination were so much less serious than the risks of variolation, but for health authorities and governments it could still be tricky. When smallpox was brought to

New York City in 1947 by a traveller from Mexico, there were twelve cases of the disease and two deaths from it. Emergency vaccination of about 6 million people within a month – itself an astonishing achievement – was successful in stopping the epidemic, and in this way doubtless saved many lives, but six people died from reactions to the vaccine.[39]

In New York in 1947 the risk of an epidemic was very real. The situation in New York or London or many other places now is that though there is a theoretical risk of a terrorist attack using smallpox it is very difficult to quantify. The risks of mass vaccination are also very small but they are fairly well known.[40] Two surveys in the United States in 1968 – when vaccination was still general – showed that, for primary vaccinations, there were between one and two deaths per million vaccinated, and adverse effects (some of them serious) in far more.* Governments wish neither to leave their populations unprotected nor to take protective measures that may turn out to have been unnecessary and that, on a country-wide scale, are likely to kill or seriously harm many individuals. The danger of a major epidemic of smallpox would need to be very high for the United States government to start a country-wide mass vaccination programme that might kill a few hundred people even if attempts were made to exclude those with eczema or with compromised immune systems.[41] So the wise short-term policy for

---

*Of the three main life-threatening complications, a post-vaccinial encephalitis occurred in 3–12 cases per million primary vaccinations, a progressive spread of the vaccinial infection continuing for many weeks occurred in 1–2 cases, and a generalised spread in patients (particularly children) with eczema occurred in 10–40 cases. Inadvertent infection of other parts of the body from the lesion at the vaccinated site occurred in about 500 cases per million vaccinations; it sometimes caused scarring but was otherwise not serious unless the eye was affected. Minor complications included a mild spread of the rash and a mild fever; in very young children up to 70 per cent were likely to suffer from fever for a day or more. The progressive and persistent spread of the vaccinial infection and the generalised spread in patients with eczema are both thought to be the result of inadequate production of antibodies to vaccinia, and patients with either of these complications can be helped by injecting them with 'vaccinia immune globulin' (VIG) – a solution of antibodies prepared from the blood plasma of *normal* recently vaccinated subjects. Plans to cope with bioterrorism therefore include the provision of adequate stocks of VIG.

any government is to ensure that it is in a position to vaccinate massively – either locally or country-wide – should the risk of smallpox become much greater, but at present to vaccinate only those who would be immediately most at risk. This seems to be the policy currently adopted by America, Britain and some other countries, with the Americans, as the most threatened, taking the most vigorous line on vaccination.

In December 2002 both the UK and the USA announced their plans. The UK planned to vaccinate about 350 health workers and to organise them into twelve regional response groups capable of 'ring-vaccinating' people around any local outbreak.[42] The US plan was on a grander scale and in three stages.[43] In the first stage there would be mandatory vaccination of about half a million selected military personnel, followed by voluntary vaccination of nearly as many medical and health-care workers. In the second stage, vaccination would be offered to police and firemen and to further medical staff – about 10 million in all. Finally – perhaps in late 2003 or 2004 – vaccination would be made available to the general public.

Vaccination of the military went ahead without delay and, thanks to the careful screening of those to be vaccinated, caused no deaths and fewer than predicted of the expected complications; unexpectedly, though, ten of the 240,000 who were being vaccinated for the first time developed inflammation of the heart or of its sheath.[44] Civilian medical and health-care workers were reluctant to be vaccinated, and by April 2003 the slow flow of volunteers had shrunk to a trickle. Even by the end of June fewer than 40,000 had been vaccinated, and in five of those vaccination had been followed by angina or by heart attacks, two of them fatal. Although four of the five had features predisposing to heart attacks, the timing suggested that vaccination might have acted as a trigger, and on 26 March the CDC recommended that persons with heart disease should not be vaccinated. Dr Raymond Strikas, the 'director of smallpox preparedness' in the CDC's immunisation division, points out that the well-publicised heart problems were not the only reason for the shortage of volunteers.[45] There was also the quick victory in Iraq, which was interpreted as a reduction in the threat to the USA, and the compensation culture in the USA, which led some groups of health-care workers to resist vaccination until there was a law to compensate them if they suffered side-effects. Whatever the reasons, by October 2003, US federal health officials accepted that the US smallpox vaccination programme had stopped.[46] This made it all the more important for governments to be able to act effectively if there were an outbreak, and there have been a number of national and international exercises simulating attacks by self-infected terrorists who deliberately spread the infection by mingling with crowds and travelling by rail and air.[47]

Stocks of vaccine in the USA proved unexpectedly useful in June and July 2003, when more than seventy human cases of suspected monkeypox were reported in six mid-western states.[48] The disease seems to have been brought to the USA by Gambian giant rats shipped from Ghana by a Texan wildlife importer and sold to an Illinois pet store, where they infected pet prairie dogs. About a quarter of the human cases were admitted to hospital and two children were seriously ill, but all recovered. Because, in the words of the deputy director of the CDC, 'the risk–benefit ratio of the vaccine' had changed, vaccination was recommended for all – even including pregnant women and children – who had had close contact with people or animals confirmed to have monkeypox. In the event, though, only thirty people were vaccinated. Judging from studies in Zaïre in the 1980s, vaccination is only about 85 per cent effective in conferring immunity to monkeypox,[49] but the risk of an uncontrolled epidemic was not too alarming because monkeypox spreads rather feebly from human to human.

Given the possible existence of smallpox stocks for many years to come, the hope is that the problem of balancing the risks and benefits of vaccination can be made easier by developing safer vaccines. Here the difficulty may be less in the development than in testing the result. There already exists a modified form of the Ankara strain of the vaccinia virus, which was prepared from the original Ankara strain, towards the end of the intensified smallpox eradication programme, by more than 500 successive passages in cultures of cells taken from chicken embryos.[50] (The thinking behind such a tedious procedure is that repeated growth in cells taken from a chicken – an abnormal host – would lead to the selection of mutant strains that thrived in chicken cells but were less suited to mammalian cells.) The resulting virus was found to have about 14 per cent less DNA, and lacked – we now know – some of the genes coding for proteins that interfere with the host cell's defences. It was also incapable of multiplying vigorously in human cells, and so could not spread through the body or infect others; but even the small amount injected was able to induce an immune response. In Germany this 'attenuated' virus was used to vaccinate more than 120,000 subjects, without any side-effects; but with no satisfactory animal model there was no acceptable way of testing its ability to prevent smallpox. Research on this virus was dramatically boosted in February 2003 by the bioterrorism-conscious United States Department of Health and Human Services, who offered contracts totalling up to $20 million for the first year of funding.[51] (This is just twelve years after the WHO, to save costs of refrigerated storage, had optimistically decided there was no longer need for smallpox vaccine, and had reduced their Geneva stockpile from 200 million doses to 500,000.)[52]

Though the eradication of smallpox has been a powerful stimulus to bioterrorism, the threat has in turn led to novel work on vaccinia; Satan is not calling all the shots. The reason why vaccinia virus is so good at immunising us against smallpox is that it is very good at inducing our immune systems to make antibodies (and killer T-cells) that recognise proteins from the vaccinia virus; and those proteins are very similar to the proteins from the closely related variola virus. Because of this similarity, the antibodies and killer cells fail to distinguish between the two viruses, so having been immunised with vaccinia we are immune to smallpox. In the 1980s a number of people had the idea of using techniques of modern molecular biology to insert, into the genome of the vaccinia virus, genes that code for some of the proteins of other disease-causing viruses – hepatitis B, influenza, herpes and rabies.[53] This was genetic engineering to prevent disease, not, as in the Russian version, to invent new horrors. The hope was that immunisation with the 'recombinant' virus – to use current jargon – would confer immunity to the disease caused by the virus that was the source of the inserted gene. And it worked. A strain of vaccinia virus, whose genome contained the gene for a surface protein of the rabies virus, has even been used to immunise foxes and racoons against rabies, by putting it on bits of meat used as bait.[54]

More recently this field of research has developed in two ways. First, immunologists have been making recombinant viruses safer by starting with attenuated vaccinia viruses, or even with avian pox viruses such as canarypox. Because these viruses replicate poorly, if at all, in humans they can be used in patients with damaged immune systems, and the hope is that they will lead to a satisfactory vaccine against the virus that causes AIDS.[55] But why stop at viral infections? In the last few years immunologists have been exploring the use of attenuated recombinant vaccinia viruses to protect against tuberculosis (a bacterial infection) and malaria (a protozoal infection).[56] And why stop at infections? Although cancer cells contain the same genes as normal cells, some of these genes are expressed only in the cells of particular cancers, so these cells contain peculiar and characteristic proteins. It is therefore possible to design recombinant vaccines that lead to the formation of antibodies and killer T-cells selective for the cells of these cancers.[57] Such vaccines, based on the modified Ankara strain of the vaccinia virus and designed for treating some cancers of the lung, breast, prostate gland or skin (melanomas) are already being tested in patients. Vaccinia viruses may have a doubtful provenance but they are certainly fashionable.

The final success of the smallpox eradication scheme was to make people think seriously about eradicating other infections. Two obvious candidates are measles and polio, both sharing with smallpox the crucial features that the virus responsible has no animal reservoir and that there is a vaccine which gives lasting immunity. In 1970 measles killed nearly 8 million world-wide; by January 2003 that figure was down to about 800,000 – though the CDC points out that 'measles remains the leading cause of vaccine-preventable childhood mortality'.[58] With polio, the risk of paralysis is much greater than the risk of death and not much less frightening. In 1980 the disease was reckoned to paralyse about half a million children a year. In an echo of the smallpox campaign, in 1985 the Pan American Health Organisation resolved to eradicate polio from the Americas, and in 1988 the WHO resolved to eradicate it from the world, setting a target date of 2000. As well as political and financial difficulties there have been unexpected technical problems. People with impaired immune systems can carry the virus for long periods and transmit it to others. And in the vaccine universally used, the attenuated form of one wild strain of polio virus is so little different from its wild parent that, very rarely, it can mutate back to a form that causes paralysis. But although the target was not met (and Henderson is said to have recommended eradicating the word 'eradication' and continuing with routine vaccination),[59] the number of estimated cases has dropped from about 350,000 in over 125 countries in 1988 to under 700 in six countries in 2003 — a fall of well over 99 per cent. Jenner would have been delighted.

And perhaps not altogether surprised. Two hundred years ago he wrote to a friend, 'The phenomena of the cow-pox open many paths for speculation, every one of which I hope may be explored.'[60]

Notes

Abbreviations

ALIBEK
Alibek, Ken (1999) *Biohazard*, Random House, New York, and Hutchinson, London; also (2000) Arrow Books, London.

BARON
Baron, John (1827, vol. I; 1838, vol. II) *The Life of Edward Jenner*, Henry Colburn, London.

BERCE
Bercé, Yves-Marie (1984) *Le chaudron et la lancette*, Presses de la Renaissance, Paris.

BRILLIANT
Brilliant, L. B. (1985) *The Management of Smallpox Eradication in India: a case study and analysis*, University of Michigan Press, Ann Arbor.

DIXON
Dixon, C. W. (1962) *Smallpox*, Churchill, London.

DNB
Dictionary of National Biography, Smith Elder, London.

EDWARDES
Edwardes, E. J. (1902) *A Concise History of Small-Pox and Vaccination in Europe*, H. K. Lewis, London.

FENNER et al.
Fenner, F., Henderson, D. A., Arita, I., Jezek, Z. and Ladnyi, I. D. (1988) *Smallpox and its Eradication*, WHO, Geneva.

HOPKINS
Hopkins, Donald R. (1983) *Princes and Peasants: Smallpox in History*, University of Chicago Press.

JENNER'S 'INQUIRY'
Jenner, Edward (1798) 'An Inquiry into the Causes and Effects of the Variola Vaccinae', London.

MILLER
Miller, Genevieve (1957) *The Adoption of Inoculation for Smallpox in England and France*, University of Pennsylvania Press.

MOORE (1815)	Moore, James C. (1815) *The History of the Small Pox*, Longman, Hurst, Rees, Orme & Brown, London.
MOORE (1817)	Moore, James C. (1817) *The History and Practice of Vaccination*, J. Callow, London.
NEEDHAM et al.	Needham, J., Lu, G-D. and Sivin, N. (2000) *Science and Civilisation in China*, vol. 6, part VI, Cambridge University Press.
Phil. Trans. Roy. Soc.	*Philosophical Transactions of the Royal Society.*
RHAZES	Rhazes, *A Treatise on the Smallpox and Measles*, trans. W. A. Greenhill (1848), Sydenham Society, London.
SIMON	Simon, John (1857) General Board of Health, *Papers relating to the History and Practice of Vaccination*, HMSO, London.
TUCKER	Tucker, J. B. (2001) *Scourge: the once and future threat of smallpox*, Atlantic Monthly Press, New York.
WHO FINAL REPORT	*The Global Eradication of Smallpox: Final Report of the Global Commission for the Certification of Smallpox Eradication* (1980), WHO, Geneva.
WONG & WU	Wong, K. C. and Wu, L-T. (1936) *History of Chinese Medicine*, 2nd edn, National Quarantine Service, Shanghai.

Chapter 1

1. Jenkins, Elizabeth (1958) *Elizabeth the Great*, Victor Gollancz, London.
2. HOPKINS, p. 74.
3. FENNER et al., p. 175.

Chapter 2

1. See article on plague in *The Jewish Encyclopedia* (1925), Funk & Wagnalls, New York.
2. MILLER, p. 103.
3. WONG & WU, pp. 269 and 273.
4. MacGowan, D. J. (1884) *Imperial Customs Medical Reports*, 27, 9.
5. Nicholas, R. W. (1981) *Journal of Asian Studies*, 41, 21–44.
6. Ibid.
7. Bhishagratna, K. L. (1963) *An English Translation of the Shushruta Samhita*, vol. 2, pp. 90 and 454, Chowkhamba Sanskrit District Office, Varanasi.
8. This is the view expressed by T. A. Wise in his *Review of the History of Medicine* (vol. 2, pp. 108–9, Churchill, London, 1867), though in his earlier *Commentary on the*

Hindu System of Medicine (pp. 233–4, Smith Elder, London, 1845) he seems to imply that a fatal form of masurika is referred to in earlier Hindu texts.

9. (i) Keene, H. G. (1908) Article on J. Z. Holwell in DNB; (ii) Nicholas, *Journal of Asian Studies*.

10. For references see Nicholas, *Journal of Asian Studies*, 41, 22.

11. (i) Nicholas, *Journal of Asian Studies*; (ii) HOPKINS, pp. 139–40 and 159–63; (iii) MOORE (1815), frontispiece.

12. Nicholas, *Journal of Asian Studies*.

13. Littman, R. J. and Littman, M. L. (1969) *Proceedings of the American Philological Association*, 100, 261–73. But cf. Holladay, A. J. and Poole, J. C. F. (1979) *Classical Quarterly*, 29, 282–300; Hornblower, S. (1991) *A Commentary on Thucydides*, vol. 1, pp. 316–27, Clarendon Press, Oxford.

14. The English translation is from Charles Forster Smith's *Thucydides: History of the Peloponnesian War, Books I and II*, 1928, Harvard University Press, Cambridge, MA.

15. Zinsser, H. (1934) *Rats, Lice and History*, pp. 119–27, Little, Brown, Boston.

16. Garrison, F. H. (1929) *An Introduction to the History of Medicine*, Saunders, Philadelphia and London. See also *Annals of Medical History*, New York, 1917–18, vol. i, 218–20.

17. Ruffer, M. A. (1919) *Mémoires de l'Institut D'Egypte*, 1, 1–8.

18. Ruffer, M. A. and Ferguson, A. R. (1911) *Journal of Pathology & Bacteriology*, 15, 1–3.

19. Ruffer, M. A. (1921) 'Pathological Notes on the Royal Mummies of a Cairo Museum', in *Studies in the Palaeopathology of Egypt* (ed. R. L. Moodie), pp. 166–78, University of Chicago Press, Chicago and London.

20. Smith, G. E. (1912) *The Royal Mummies*, Imprimerie de l'Institut Français d'Archéologie Orientale, Cairo.

21. HOPKINS, p. 15.

Chapter 3

1. (i) WONG & WU, p. 274; (ii) NEEDHAM et al., pp. 125–7.

2. (i) Littman, R. J. and Littman, M. L. (1973) *American Journal of Philology*, 94, 243–55; (ii) Zinsser, H. (1934) *Rats, Lice and History*, pp. 135–7, Little, Brown, Boston.

3. (i) *Scriptores Historiae Augustae* (trans. D. Magie, 1929) I, p. 203, Harvard University Press, Cambridge, MA; (ii) Dio Cassius, *Roman History* (trans. E. Cary, 1927), IX, p. 117, Heinemann, London.

4. (i) DIXON, p. 189; (ii) Willan, R. (1821) 'An Inquiry into the Antiquity of Small-Pox, Measles and Scarlet Fever', pp. 5–10, in *Miscellaneous Works of the late Robert Willan, M.D.* (ed. Ashby Smith), Cadell, London.

5. (i) HOPKINS, pp. 101–2; (ii) MOORE (1815), pp. 96–7.

6. (i) Margoliouth, D. S. (1911) article on Mahomet in *Encyclopaedia Britannica*, 11th edn, Cambridge University Press, Cambridge; (ii) HOPKINS, pp. 165–6; (iii) MOORE (1815), pp. 49–55. But cf. Beeston, A. F. L. (1979) article on Abraha in *The Encyclopaedia of Islam*, Brill, Leiden.

7. Willan, 'Inquiry into the Antiquity', pp. 91–2.
8. (i) RHAZES, pp. 103–5, 129–30 and 163–4; (ii) MOORE (1815), pp. 113–15.
9. Haygarth, J. (1793) 'Sketch of a plan to exterminate the casual small-pox from Great Britain', p. 41, J. Johnson, London.
10. MOORE (1815), p. 63.
11. (i) Murdoch, J. (1910) *A History of Japan*, vol. 1, pp. 105 and 110–18, Asiatic Society of Japan, Kegan Paul, Trübner, London; (ii) Fujikawa, Y. (1969, originally 1912) *Nihon Shippei-Shi (History of Epidemic Diseases in Japan)*, Tokyo; (iii) Ponsonby Fane, R. (1959) *The Imperial House of Japan*, pp. 45–7, Ponsonby Memorial Society, Kyoto.
12. Murdoch, *History of Japan*, vol. 1, p. 92.
13. (i) Fujikawa, *Nihon Shippei-Shi*; (ii) Murdoch, *History of Japan*, vol. 1, p.192.
14. Boorstin, D. (1983, paperback 2001) *The Discoverers*, p. 500, Phoenix Press, London.
15. *American Dictionary of Scientific Biography*, 1975.
16. RHAZES, p. 29.
17. MOORE (1815), p. 125.
18. NEEDHAM et al., p. 129.
19. RHAZES, p. 47.
20. Ibid., p. 103.
21. Ibid., p. 39.
22. MOORE (1815), p. 137.
23. Ibid., pp. 136–7.
24. Ibid., pp. 143–4.
25. (i) Ibid., p. 163; (ii) Garrison, F. H. (1929) *History of Medicine*, 4th edn, p. 164, Saunders, Philadelphia and London.
26. Moore, Norman (1908) Entry for Gaddesden, John of, in DNB.
27. (i) HOPKINS, pp. 28, 108, 137, and 295–300; (ii) NEEDHAM et al., p. 161; (iii) MOORE (1815), pp. 162–3.
28. Lebeuf, L. G. (1900) *Indian Lancet* (Calcutta), 15, 492–4; quoted in HOPKINS, p. 297.
29. DIXON, p. 191.
30. Porter, J. 'Queries sent to a Friend by Dr Maty & answered by his Excellency James Porter Esq', *Phil. Trans. Roy. Soc.*, XLIX, p. 108.
31. Cao Xueqin (eighteenth century) *The Story of the Stone* (trans. David Hawkes), vol. 1, p. 424, Penguin, 1973.
32. Finsen, N. R. (1901) *Phototherapy* (trans. J. H. Sequira), p.1, Arnold, London.
33. Ricketts, T. F. and Byles, J. B. (1903) *The Lancet*, 30 July, 287–90.
34. (i) EDWARDES, p. 4; (ii) MOORE (1815), pp. 94–6.
35. Edwardes, E. J. (1902) *British Medical Journal*, ii, pp. 27–30.
36. Fisher, H. A. L. (1949) *A History of Europe*, vol. 1, p. 479, Eyre & Spottiswoode, London.

Chapter 4

1. SIMON, pp. ii–iii.
2. (i) HOPKINS, p. 205; (ii) McNeill, W. H. (1976) *Plagues and Peoples*, Penguin reprint 1994, p. 192.
3. Spanish eyewitness, quoted in HOPKINS, p. 207.
4. Fray Toribio Motolinía, quoted in HOPKINS, p. 206.
5. HOPKINS, p. 215.
6. Oldstone, M. (1998) *Viruses, Plagues, and History*, p. 32, Oxford University Press.
7. HOPKINS, p. 215.
8. (i) Ibid., p. 234; (ii) McNeill, *Plagues and Peoples*, p. 195.
9. Quoted by Mark Wheelis in *Biological and Toxin Weapons* (ed. E. Geissler and J. E. van C. Moon), p. 19, Oxford University Press.
10. Mather, Increase (1675) 'A Relation of the Troubles which have happened in New England by Reason of the Indians there', Boston; quoted in Duffy, J. (1951) 'Smallpox and the Indians in American Colonies', *Bulletin of the History of Medicine*, vol. 25.
11. Heagerty, J. J. (1928) *Four Centuries of Medical History in Canada*, vol. 1, p. 22, John Wright, Bristol.
12. HOPKINS, p. 236.
13. Stearn, E. W. and Stearn, A. E. (1945) *The Effect of Smallpox on the Destiny of the Amerindian*, Bruce Humphries, Boston.
14. Quoted in Stearn and Stearn, *Effect of Smallpox*, p. 33.
15. Father Millet (1691) quoted in Duffy 'Smallpox and the Indians'.
16. Father Chirino (1591) quoted in HOPKINS, p. 114.
17. *Chronicles of the Kings of Tripura*, quoted in HOPKINS, p. 142.
18. Quoted in HOPKINS, p. 143.
19. Fawcett, C. (1936) *The English Factories in India*, Oxford University Press.
20. Letter from Matthew Hale quoted in 'Small-Pox before Jenner', *British Medical Journal*, Jenner Centenary Number, 23 May 1896, p. 1262.
21. *London Gazette*, 23 February 1688, quoted in 'Small-Pox before Jenner'.
22. Pepys, Samuel, *Diary*, 9 February 1668.
23. Ibid., 30 March 1668.
24. Evelyn, John, *Diary*, 29 December 1694.
25. Keynes, W. M. (1997) *Journal of the Royal Society of Medicine*, January, p. 60.
26. Creighton, Charles (1894) *A History of Epidemics in Britain*, Cambridge University Press; figures taken from the London Bills of Mortality.
27. Wrightson, K. (1982) *English Society 1580–1680*, Hutchinson, London.
28. Singer, C. and Singer, D. (1913) 'The Development of the Doctrine of Contagium Vivum, 1500–1750', *XVIIth International Congress of Medicine*, section 23, pp. 187–8.
29. Latham, R. G. (1848–50) *The Works of Thomas Sydenham translated from the Latin ...*, vol. I, Sydenham Society, London.
30. MOORE (1815), pp. 137–8.
31. Nicholas, R. W. (1981) *Journal of Asian Studies*, 41, 21–44.

32. Dover, Thomas (1733) *The Ancient Physician's Legacy to his Country*, London.
33. Mead, Richard (1748) *Discourse on Small Pox and Measles*, London.

Chapter 5

1. Woodward, J. (1713) 'Extract of a letter from Emanuel Timonius', *Phil. Trans. Roy. Soc.*, XXIX, 72–82.
2. Russell, P. (1768) 'An Account of Inoculation in Arabia', *Phil. Trans. Roy. Soc.*, LVIII, 140 50.
3. Maitland, C. (1722) 'Account of Inoculating the Smallpox', London.
4. Grundy, I. (1999) *Lady Mary Wortley Montagu*, p. 258, Oxford University Press, Oxford.
5. Evelyn, John, *Diary*, 15 September 1685.
6. Haygarth, John (1784) 'An Inquiry How to Prevent the Smallpox', London.
7. HOPKINS, p. 270.
8. Russell, 'Account of Inoculation'.
9. Maty, M. (1768) 'A Short Account of the Manner of inoculating the Small Pox on the Coast of Barbary', *Phil. Trans. Roy. Soc.*, LVIII, 128–31.
10. Creighton, C. (1894) *A History of Epidemics in Britain*, Cambridge University Press.
11. Williams, Perrot (1723) 'A Method of procuring the Small Pox, used in South Wales', *Phil. Trans. Roy. Soc.*, XXXII, 262–9.
12. (i) WONG & WU, p. 215; (ii) Chia-Feng Chang (1996) Aspects of smallpox and its significance in Chinese history, PhD thesis for London University.
13. NEEDHAM et al., pp. 131 and 143.
14. Waterson, A. P. and Wilkinson, L. (1978) *An Introduction to the History of Virology*, pp. 56–9. We are grateful to G. L. Smith for pointing out the analogy.
15. Hookham, H. (1969) *A Short History of China*, New American Library, New York.
16. Quoted in NEEDHAM et al., p. 140.
17. Nicholas, R. W. (1981) *Journal of Asian Studies*, 41, 21–44.
18. Ibid.
19. MILLER, pp. 48–9.
20. Silverman, K. (1984) *The Life and Times of Cotton Mather*, pp. 338–9, Harper & Row, New York.
21. Sloane, Hans (1756, but written in 1736) 'An Account of Inoculation', *Phil. Trans. Roy. Soc.*, XLIX (Pt II), 516–20.
22. MILLER, pp. 80–81.
23. Vanbrugh, J. (1721) Letter of 16 November; *The Complete Works* (ed. B. Dobrée and G. Webb, 1928), vol. 4, Bloomsbury, London.
24. *Applebee's Original Weekly Journal*, August 1721, quoted in MILLER, p. 87.
25. Sloane, 'An Account'.

Chapter 6

1. *Cleveland Press*, 23 January 1957, quoted in Miller, G. (1981) *Bulletin of the History of Medicine*, 55, 9.

2. Letter 23 March 1718, in *Embassy to Constantinople: the Travels of Lady Mary Wortley Montagu*, (ed. C. Pick, 1988), p. 176, Century, London.

3. Arbuthnot, John (1722) 'Mr Maitland's Account of Inoculating the Smallpox Vindicated', London.

4. Letter August 1721, quoted in E. St John Brooks (1954) *Sir Hans Sloane*, Batchworth Press, London.

5. Letter August 1722 quoted in St John Brooks, *Sir Hans Sloane*.

6. FENNER et al., p. 254.

7. Bulloch, J. M. (1930) *A Pioneer of Inoculation; Charles Maitland*, Aberdeen University Press.

8. Silverman, K. (1984) *The Life and Times of Cotton Mather*, p. 343, Harper Row, New York.

9. Mather, Cotton (1722) *An Account of the Method and Success of Inoculating the Small-Pox, in Boston in New England*, London.

10. Douglass to Colden, 1 May 1722, quoted in Creighton, Charles (1894) *A History of Epidemics in Britain*, II, p. 607, Cambridge University Press.

11. Newman, Henry (1722) *Phil. Trans. Roy. Soc.*, XXXII, 33–5.

12. Silverman, *Cotton Mather*, p. 349.

13. Mather, p. 22.

14. Barrett, J. J. (1942) 'The Inoculation Controversy in Puritan New England', *Bulletin of the History of Medicine*, vol. 12.

15. Massey, Edmund (1722) 'A Sermon against the Dangerous and Sinful Practice of Inoculation', London.

16. Wagstaffe, W. (1722) 'A Letter to Dr Freind; Shewing the Danger and Uncertainty of Inoculating the Small Pox', London.

17. Sparham, Legard (1722) 'Reasons against the Practice of Inoculating the Small-Pox', London.

18. Arbuthnot, 'Mr Maitland's Account'.

19. Maty, M. (1755) *Phil. Trans. Roy. Soc.*, XLIX, 96–109.

20. Newspaper advertisement, 1774, quoted in *British Medical Journal*, Jenner Centenary Number, 23 May 1896, p. 1261.

21. Nettleton, T. (1722) *Phil. Trans. Roy. Soc.*, XXXII, 370, 35–48.

22. Nettleton to Jurin, 24 January 1723, quoted in Rusnock, A. (2002) ' "The Merchant's Logick": Numerical Debates over Smallpox Inoculation in 18th Century England', in *The Road to Medical Statistics* (ed. E. Magnello and A. Hardy), vol. 67 in the Wellcome Series in the History of Medicine, pp. 37–54, Clio Medica, Amsterdam.

23. Jurin, J. (1722) *Phil. Trans. Roy. Soc.*, XXXII, 213–27.

24. Jurin, J. (1724) 'An Account of the Success of Inoculating the Small-Pox in Great Britain', London.

25. Jurin, *Phil. Trans. Roy. Soc.*, XXXII, 213–27.

26. Vanbrugh, Sir John, *The Complete Works* (1928) (ed. B. Dobrée and G. Webb), Bloomsbury.

27. Letter to Sloane dated 15 June 1723, quoted in MILLER, p. 190.

28. Bernoulli, D. (1760) *Histoire de l'Académie Royale des Sciences 1760* (Paris, 1766), Pt II, 1–45. For English translation and comments see Bradley, L. (1971) *Smallpox Inoculation: an eighteenth-century mathematical controversy*, University of Nottingham, Adult Education Department.

29. Franklin, Benjamin (1759) in 'Preface to Dr Heberden's Pamphlet on Inoculation', London.

30. Letter from Josiah Wedgwood to Thomas Bentley, 2 March 1767, quoted in Uglow, Jenny (2002) *The Lunar Men*, p. 175, Faber & Faber, London.

31. Madame D'Arblay (Fanny Burney) to Dr Burney, 16 March 1797, *Diary and Letters of Madame D'Arblay*, (1905), vol. 5, p. 321, Macmillan, London.

32. Woodforde, James, *The Diary of a Country Parson*, entry for 7 March 1791.

33. Franklin, Benjamin, *Autobiography*, written 1771–89.

Chapter 7

1. Schultz, David (1758) 'An Account of Inoculation', London.

2. Montagu, Lady Mary Wortley, 'An Account of the Inoculating the Small Pox at Constantinople', published anonymously as 'by a Turkey-Merchant', in *The Flying-Post*, September 1722.

3. Fiske, Dorothy (1959) *Dr Jenner of Berkeley*, Heinemann, London.

4. HOPKINS, p. 60.

5. Kirkpatrick, J. (1743) *An Essay on Inoculation*, London.

6. Creighton, C. (1894) *A History of Epidemics in Britain*, Cambridge University Press.

7. *Gentleman's Magazine*, 1765, p. 163.

8. Quoted in Janssens, U. (1981) 'Matthieu Maty and the adoption of inoculation for smallpox in Holland', *Bulletin of the History of Medicine*, 55, 256.

9. The phrase was used by Bamber Gaskoyne (1725–91), see Dobson, M. J. (1997) *Countours of Death and Disease in Early Modern England*, Cambridge University Press.

10. Letter of 26 February 1769, quoted in Philip Clendenning, 'Dr Thomas Dimsdale and Smallpox Inoculation in Russia', *Journal of the History of Medicine*, XXVIII, April 1973.

11. Klebs, A. C. (1914), quoted in HOPKINS p. 69.

12. Dimsdale, T. *Tracts*, St Petersburg, quoted in *Papers relating to the History and Practice of Vaccination* (1857), HMSO, London.

13. Abraham, J. J. (1933) *Lettsom*, p. 204, Heinemann, London.

14. Clendenning, 'Smallpox Inoculation in Russia'.

15. Letter to Lorenz Hagenauer, 22 February 1764; in *The Letters of Mozart and his Family* (1985), (ed. Emily Anderson), Macmillan, London.

16. Ibid., 10 November 1767.

17. Edwardes, E. J. (1902) *British Medical Journal*, ii, 27–30.

18. Mead, Richard (1748) *Discourse on Small Pox and Measles*, p. 11, London.

19. Laidler, P. W. and Gelfand, M. (1971) *South Africa: Its Medical History 1652–1898*, C. Struik, Cape Town.

20. Burton, R. (1860, reprinted 1961) *The Lake Regions of Central Africa*, vol. I, p. 166, Horizon Press, New York.
21. Gelfand, M. (1957) *Livingstone the Doctor*, p. 138, Blackwell, Oxford.
22. Stanley, H. (1899, reprinted 1988) *Through the Dark Continent*, vol. II, p. 49, Dover Publications, New York.
23. Humboldt, A. von, (1821) *Travels 1794–1804*, vol. 5, p. 28, Longman, London.
24. Bruce, James (1790) *Travels to Discover the Source of the Nile*, Edinburgh University Press, 1964.
25. *Journal of the Royal Asiatic Society*, XXIII, 1794.
26. *British Medical Journal*, Jenner Centenary Number, 23 May 1896, p. 1264.
27. HOPKINS, p. 146.
28. Campbell, J. (2002) *Invisible Invaders*, Melbourne University Press.
29. HOPKINS, p. 220.
30. Heagerty, J. J. (1928) *Four Centuries of Medical History in Canada*, vol. I, p. 75, John Wright, Bristol.
31. Anderson, F. (2000) *Crucible of War*, p. 199, Alfred A. Knopf, New York.
32. HOPKINS, p. 246.
33. Quoted by Mark Wheeler in Geissler, E. and Moon, J. E. van C. (1999) *Biological & Toxin Weapons*, p. 22, Oxford University Press, Oxford.
34. Marble, A. E. (1993) *Surgeons, Smallpox and the Poor*, McGill-Queen's University Press, Montreal.
35. *Gazette du Canada*, 11 April 1765, quoted in Heagerty, *Four Centuries*, vol. I, p. 77.
36. Simpson, H. N. (1954) *New England Journal of Medicine*, 250, no. 16, 681.
37. Thursfield, H. (1940) *Annals of Medical History*, pp. 312–18.
38. Ibid.
39. Letter from Hannah Winthrop to Mercy Warren, quoted in Blake, John (1959) *Public Health in the Town of Boston 1630–1822*, Harvard University Press.
40. HOPKINS, p. 255.
41. Fenn, E. A. (2001) *Pox Americana*, pp. 59 and 127, Hill & Wang, New York.
42. (i) Thursfield, *Annals*, pp. 312–18; (ii) HOPKINS, p. 260.
43. Sabin, Josiah, quoted in Dann, J. D. (ed.) (1980) *The Revolution Remembered*, p. 19, University of Chicago Press.
44. Fenn, *Pox Americana*, p. 68.
45. Letter from George Washington to Dr Shippen, January 1777, quoted in Thursfield, *Annals*, pp. 312–18.
46. Trevelyan, G. O. (1912) *George III and Charles James Fox*, Longmans Green, London.
47. Fenn, E. *Pox Americana*, p. 122.
48. Boswell, James (1786) *Journal of a Tour in the Hebrides*.
49. Creighton, C. (1894) *A History of Epidemics in Britain*, vol. 2, p. 517, Cambridge University Press.
50. Figures estimated by E. A. Wrigley, quoted in Inwood, S. (1998) *A History of London*, Macmillan, London.

51. Bose, K. C., *Indian Medical Gazette* (1890). The phrase is the source of the title of D. R. Hopkins' 1983 book.

52. SIMON, p. ix.

53. Quoted in Creighton, *History of Epidemics*, vol. 2, p. 507.

54. Haygarth, John (1793) 'Sketch of a plan to exterminate the casual smallpox from Great Britain and to introduce general inoculation', London.

55. Peter Razzell (1977), in attempting to show how inoculation reduced the smallpox death rate in the eighteenth century, cites figures of general inoculation in relatively small communities, where it could certainly be successful. See his *The Conquest of Smallpox*, Chapter 9, Caliban Books, Sussex.

56. Woodforde, James, *The Diary of a Country Parson*, 7 March 1791.

57. Cole, G. D. H. (1938) *Persons and Periods*, p. 66, Macmillan, London.

Chapter 8

1. BARON, vol. II, p. 168.

2. Humphry Davy, quoted by F. D. Drewitt (1931) in an introduction to *The Note-book of Edward Jenner*, p. 5, Oxford University Press, Oxford.

3. MOORE (1817).

4. William Hunter's prospectus, quoted in Fisk, Dorothy (1959) *Dr Jenner of Berkeley*, Heinemann, London.

5. BARON, vol. I, pp. 15–16.

6. *Letters from the past, from John Hunter to Edward Jenner*, Royal College of Surgeons of England (1976), London.

7. The article, published in 1892, was by Norman Moore (sometime President of the Royal College of Physicians).

8. BARON, vol. I, p. 48.

9. Letter quoted in Pearson, George (1798) 'An inquiry concerning the history of the cowpox', London.

10. Wilkinson, L. (1979) 'Smallpox and the evolution of ideas on acute (viral) infections', *Medical History*, 23, 11.

11. (i) NEEDHAM et al., p. 152; (ii) WONG & WU, p. 216.

12. (i) 'A Century of Vaccination', *British Medical Journal*, Jenner Centenary Number, 23 May 1896, p. 1265; (ii) EDWARDES, pp. 33–4.

13. Lettsom, J. C. (1801) *Observations on the Cow Pock*, London, quoted in Crookshank, E. (1889) *History and Pathology of Vaccination*, p. 105, H. K. Lewis, London.

14. JENNER'S 'INQUIRY', p. 7.

15. *Letters from the past*, p. 9.

16. Mather, Cotton (1722) *An Account of the Method and Success of Inoculating the Small-Pox, in Boston in New England*, p. 8, London, quoted in MILLER, p. 107.

17. Letter at Royal College of Surgeons, quoted in Fisher, R. (1991) *Edward Jenner*, p. 67, André Deutsch, London.

18. Letter to John Ring, 16 August 1799, quoted in BARON, vol. I, p. 356.

19. BARON, vol. II, p. 168.

20. Letter from Home to Banks, 22 April 1897, quoted in Baxby, D. (1999) *Medical History*, 43, 108–10.
21. JENNER'S 'INQUIRY', p. 24.
22. Ibid., p. 33.
23. Fisher, *Jenner*, p. 75.
24. Letter 9 August 1798, in Jenner, E. (1799) *Further Observations on the Variolae Vaccinae or Cow Pox*, London.
25. Letter dated 12 October 1798, quoted in Crookshank, *History and Pathology*, p. 145.
26. In his *Further Observations*, Jenner described four distinct types of spurious cowpox.
27. BARON, vol. II, p. 135.
28. Woodville, William (1799) *Reports of a Series of Inoculations for the Variolae Vaccinae*, London.
29. Woodville, William (1800) *Observations on the Cow-pox*, London.
30. BERCE, p. 174.
31. BARON, vol. II, p. 238. In 1813 the Smallpox Hospital reported that 100,000 persons had been 'protected and relieved' since the hospital was founded in 1746 – see Hutchinson, J. R. (1947) *A Historical Note on the Prevention of Smallpox in England*, Ministry of Health, UK.
32. Quoted in Anderson, E. G. 'The History and Effects of Vaccination', *Edinburgh Review*, April 1899, p. 346. (Fishtail gas lights were invented in 1820.)
33. Moore, Norman, article on Moseley in DNB (1894).
34. (i) Rowley, William (1805) *Cow-Pox Inoculation No Security against Small-Pox Infection*, London; (ii) Thornton, R. J. (1806) *Vaccinae Vindicia*, London, quoted in Fisher, *Jenner*, p. 165.
35. Birch, John (1805) 'A Letter occasioned by the many failures of Cow-Pox', London, reprinted in Crookshank, *History and Pathology*, vol. II, London.
36. Birch, John (1806) *Serious Reasons for Uniformly Objecting to the Practice of Vaccination*, London, quoted in DIXON, p. 287.
37. Malthus, Thomas (1798) *Essay on Population*, quoted in Shurkin, Joel (1979, reprinted 2000) *The Invisible Fire*, pp. 182–4, Authors Guild, USA.
38. See *Edinburgh Review*, 1806, vol. IX, 32–66.
39. 'Observations on Vaccine Inoculation', *The Medical and Physical Journal*, 1802, p. 169, quoted in Fisher, *Jenner*.
40. Letter, 1801, in BARON, vol. I, p. 326.
41. Letter, 19 April 1803, in BARON, vol. I, p. 596.
42. BARON, vol. II, p. 192.
43. Ibid., vol. I, p. 541.

Chapter 9

1. Sigerist, Henry (1950) 'Letters of Jean de Carro to Alexandre Marcet', *Supplement to the Bulletin of the History of Medicine*, no. 12, Johns Hopkins Press, Baltimore.
2. BERCE, p. 61.
3. Fisher, R. B. (1991) *Edward Jenner*, p. 109, André Deutsch, London.

4. *Letters of Mary Nisbet of Dirleton, Countess of Elgin* (ed. J. P. Nisbet Hamilton Grant, 1926), p. 110, John Murray, London.
5. HOPKINS, p. 149.
6. 'Diffusion of Vaccination', *British Medical Journal*, Jenner Centenary Number, 23 May 1896, p. 1267.
7. BARON, vol. I, p. 410.
8. Royal order, 1 September 1803, quoted in HOPKINS, p. 224.
9. Smith, Michael (1974) 'The "Real Expedición Marítima de la Vacuna" in New Spain and Guatemala', *Transactions of the American Philosophical Society*, vol. 64, part I, pp. 1–74.
10. Letter from Jenner to Richard Phillips, 16 January 1807, in an archive at the Royal College of Physicians (quoted in Fisher, *Jenner*, pp. 179–80).
11. Bowers, J. Z. (1981) 'The Odyssey of Smallpox Vaccination', *Bulletin of the History of Medicine*, 55, 30.
12. Babeau, Albert (1892) *Paris en 1789*, Paris, quoted in Bazin, H. (2000) *The Eradication of Smallpox*, p. 87, Academic Press, London.
13. BERCE, p. 116.
14. Letter from Jenner to the National Institute of France, 1803, see BARON, vol. I, p. 603.
15. Meynell, Elinor (1987) 'Thomas Michael Nowell and his "matière de Boulogne"', *Journal of the Royal Society of Medicine*, 80, 232–7.
16. MOORE (1817).
17. Ibid.
18. Letter from Marshall to Jenner, 26 January 1802, quoted in BARON, vol. I, p. 403.
19. Letter from Luigi Sacco to Jenner, 5 January 1808, quoted in BARON, vol. II, p. 112.
20. BARON, vol. I, p. 265.
21. General Council of the Department of Indre and Loire, in *Report of Comité Central de Vaccine 1803*, quoted in Bazin, *Eradication*, p. 90.
22. Gérin, Winifred (1981) *Horatia Nelson*, p. 7, Oxford University Press.
23. Figures from EDWARDES, pp. 45–6.
24. Laidler, P. W. and Gelfand, M. (1971) *South Africa: Its Medical History 1652–1898*, C. Struik, Cape Town.
25. Revd Canon Lockyer in *The Telegram*, quoted in Heagerty, J. J. (1928) *Four Centuries of Medical History in Canada*, vol. I, p. 84, John Wright, Bristol.
26. Quoted in Blake, J. B. (1957) *Benjamin Waterhouse and the Introduction of Vaccination*, pp. 46–7, University of Pennsylvania Press, Philadelphia.
27. Blake, J. B. (1959) *Public Health in the Town of Boston 1630–1822*, Harvard University Press, Cambridge, MA.
28. Shurkin, J. N. (1979, reprinted 2000) *The Invisible Fire*, p. 193, Authors Guild.
29. Letter from Thomas Jefferson to Jenner, 14 May 1806, quoted in BARON, vol. II, p. 94.
30. WONG & WU, p. 277.
31. Alexander Pearson's *Report submitted to the Board of the National Vaccine Establishment*, 1816, quoted in WONG & WU.

32. Fisk, Dorothy (1959) *Dr Jenner of Berkeley*, p. 266, Heinemann, London.
33. Jenner, E. (1801) *The Origin of the Vaccine Inoculation*, London.

Chapter 10
1. The full text is in SIMON, Appendix D.
2. Creighton, C. (1894) (quoting John Cross) *A History of Epidemics in Britain*, Cambridge University Press.
3. Cross, John, *History of the Variolous Epidemic at Norwich, in the year 1819*, quoted in SIMON, p. 120.
4. Marson's 1856 analysis is in SIMON, p. lxxiv.
5. SIMON, p. xxvi.
6. Ibid., pp. 128 and 155.
7. Le Fanu, W. R. (1951) *A Bio-bibliography of Edward Jenner*, Harvey & Blythe, London.
8. Letter to Dunning, 25 October 1804, quoted in BARON, vol. II, p. 344.
9. Sloane, H. (1736, published 1756), 'An Account of Inoculation', *Phil. Trans. Roy. Soc.*, XLIX, 519.
10. DIXON, p. 142.
11. BARON, vol. I, p. 157.
12. Goldson, W. (1804) *Some Recent Cases of Small Pox subsequent to Vaccination ...*, William Woodward, Portsea.
13. Ring, J. (1804) *An Answer to Mr Goldson, proving that Vaccination is a Permanent Security against the Small-Pox*, Murray, London.
14. Article by D'Arcy Power on *John Ring (1752–1821)* in DNB.
15. 'A Century of Vaccination', *British Medical Journal*, Jenner Centenary Number, 23 May 1896, p. 1266.
16. EDWARDES, p. 134.
17. BERCE, p. 92.
18. De Carro, 5 June 1802, quoted in BERCE, p. 75.
19. Letter to Dunning, 21 February 1806, quoted in BARON, vol. II, p. 351.
20. BARON, vol. II, p. 69.
21. Letter from Jenner to Lettsom, July 1807, quoted in Abraham, J. J. (1933) *Lettsom*, p. 355, Heinemann, London.
22. Report of Royal College of Physicians to Parliament, 8 July 1807.
23. Baxby, D. (1999) 'The End of Smallpox', *History Today*, March.
24. *The Lancet*, 6 June 1840.
25. SIMON, p. lxix.
26. *The Lancet*, 18 June 1853, p. 564.
27. SIMON, p. lxxv.
28. Allison, W. P. (1857), in SIMON, p. 123.
29. BARON, vol. II, p. 377.
30. BERCE, p. 60.
31. HOPKINS, p. 275.

32. Quoted by Blake, J. B. (1959) *Public Health in the Town of Boston, 1630–1822*, p. 109, Harvard University Press, Cambridge, MA.
33. Letter to the Bristol ophthalmic surgeon John Bishop Estlin, 16 January 1839, quoted in Crookshank, E. M. (1889) *History and Pathology of Vaccination*, vol. II, p. 341, H. K. Lewis, London.
34. Catlin, G. (1841) *North American Indians*, pp. 608–9, Chatto & Windus, London.
35. Wheelis, M. (1999) in *Biological and Toxin Weapons* (ed. E. Geissler and J. E. van C. Moon), p. 28, Oxford University Press.
36. Stearn, E.W. and Stearn, A. E. (1945) *The Effect of Smallpox on the Destiny of the Amerindian*, p. 86, Bruce Humphries, Boston.
37. Wheelis, in *Weapons*, p. 26.
38. HOPKINS, pp. 228–9.
39. Morrison, G. E. (1895) *An Australian in China*, Horace Cox, London.
40. WONG & WU, p. 289.
41. SIMON, p. lxvii.

Chapter 11

1. Figures from EDWARDES, p. 73.
2. Edwardes E. J. (1902) *British Medical Journal*, 5 July, p. 27.
3. McVail, J. C. (1887) *Vaccination Vindicated*, p. 167, Cassell, London.
4. Fujikawa, Y. (1969, originally 1912) *Nihon Shippei-Shi (History of Epidemic Diseases in Japan)*, Tokyo.
5. (i) 'Diffusion of Vaccination', *British Medical Journal*, Jenner Centenary Number, 23 May 1896, p. 1267; (ii) Bazin, H. (2000) *The Eradication of Smallpox*, Academic Press, London.
6. Parker, E. H. (1907) *British Medical Journal*, 12 January, p. 88.
7. Morrison, G. E. (1895) *An Australian in China*, Horace Cox, London.
8. FENNER et al., pp. 1259–60. See also Chapter 14, p. 197.
9. Das, S.C. (1902) *Journey to Lhasa and Central Tibet*, p. 193, John Murray, London.
10. 'Small-pox before Jenner', *British Medical Journal*, Jenner Centenary Number, 23 May 1896, p. 1264.
11. HOPKINS, pp. 127–8.
12. Ibid., p. 152.
13. Pringle, R. (1869) 'On Smallpox and Vaccination in India', *The Lancet*, 9 and 16 January.
14. 1928 report, quoted in HOPKINS, p. 155.
15. Campbell, J. (2002) *Invisible Invaders*, p. 227, Melbourne University Press.
16. Ibid., p. 62.
17. BERCE, p. 250.
18. HOPKINS, p. 279.
19. (i) Empson, J. (1996) *Journal of the Royal Society of Medicine*, September, pp. 514–18; (ii) Scally, G. and Oliver, I. (2003) *The Lancet*, 4 October, p. 1092.
20. Figures from HOPKINS, pp. 90–91.

21. Edwardes, *British Medical Journal*.
22. SIMON, p. xl.
23. NEEDHAM et al., p. 131.
24. HOPKINS, p. 289.
25. McVail, J. (1919) *Half a Century of Smallpox and Vaccination*, p. 8, Livingstone, Edinburgh.
26. DIXON, pp. 142–3.
27. Anderson, E. G. (1899) *The Edinburgh Review*, CLXXXIX, 335–68. She notes that vaccination mortality in Germany was only one in 100,000.
28. HOPKINS, p. 93, quoting Brown, J. R. and Maclean, D. M. (1962) *Canadian Medical Association Journal*, 87, 765–7.
29. Kidd, B. E. and Richards, M. E. (1933) *Hadwen of Gloucester*, p. 83, John Murray, London.
30. Case reported in *The Lancet*, 23 January 1897.
31. Laidler, P. W. and Gelfand, M. (1971) *South Africa: Its Medical History 1652–1898*, p. 432, C. Struik, Cape Town.
32. Anderson, *Edinburgh Review*.
33. Kidd and Richards, *Hadwen*, p. 106.
34. Baxby, D. (1999) 'The End of Smallpox', *History Today*, March.
35. Evidence to the Vaccination Commission, reported in the *British Medical Journal*, 23 May 1896, p. 1295.
36. Crookshank, E. M. (1889) *History and Pathology of Vaccination*, vol. I, p. 464, H. K. Lewis, London.
37. Millard, K. C. (1914) *The Vaccination Question*, p. 171, H. K. Lewis, London.
38. Figures from Bazin, H. (2000) *The Eradication of Smallpox*, p. 122, Academic Press, London.
39. Quoted in Millard, *Vaccination Question*, p. 3.
40. National Anti-Vaccination League Poster in Westminster, 1906.
41. Anderson, *Edinburgh Review*.
42. Wallace, Alfred Russel (1885) *45 Years of Registration Statistics proving Vaccination to be both useless and dangerous*, E. W. Allen, London. McVail's *Vaccination Vindicated* was written as an answer to Wallace.
43. McVail, *Half a Century*, p. 42.
44. Millard, *Vaccination Question*, p. 125.
45. FENNER et al., p. 65.
46. Manuscript letter from Erasmus Darwin to his son Robert, 4 September 1791, University Library, Cambridge (DAR 227.1.139).
47. Simpson, J. Y. (1868) *Proposal to stamp out smallpox*, Edmonton & Douglas, Edinburgh.
48. Introduction to *Report on an Epidemic of Small-Pox at Sheffield during 1887–88*, HMSO, 1889.
49. Letter 30 May 1889, quoted in Ramm, A. (1990) *Beloved and Darling Child*, Alan Sutton, London.

50. Cohen, M. N. (1995) *Lewis Carroll*, p. 393, Macmillan, London.

51. Burne, John (1989) *Dartford's Capital River*, Barracuda Books, London.

Chapter 12

1. Gibbs, E. P. J., Johnson, R. H. and Collings, D. F. (1973) *Veterinary Record*, 20 January, pp. 56–64.

2. Baxby, D. (1981) *Jenner's Smallpox Vaccine*, p. 170, Heinemann, London.

3. Baxby D. (1994) *Vaccination*, p. 21, Jenner Educational Trust, Berkeley, England.

4. (i) NEEDHAM et al., pp. 151–2; (ii) WONG & WU, p. 216.

5. EDWARDES, p. 33.

6. HOPKINS, p. 225.

7. BARON, vol. 2, p. 227.

8. HOPKINS, p. 268.

9. Pearson, G. (1798) *An Inquiry Concerning the History of the Cow Pox*, Johnson, London.

10. Loy, J. G. (1801) *An Account of some Experiments on the Origin of the Cow-Pox*, Whitby, England.

11. Jenner's manuscript journal, 17 May 1817, quoted in BARON, vol. 2, p. 226.

12. Crookshank, E. M. (1889) *History and Pathology of Vaccination*, vol. 1, pp. 390 ff, H. K. Lewis, London.

13. Letter to James Moore, 27 October 1813, quoted in Crookshank, *History and Pathology*, vol. 1, p. 391.

14. BARON, vol. 1, p. 238 and vol. 2, p. 335.

15. JENNER'S 'INQUIRY'.

16. Epps, J. (1831) *The Life of John Walker*, London, quoted in Fisher, R. (1991) *Edward Jenner*, p. 171, André Deutsch, London.

17. (i) Copeman, S. M. (1899) *Vaccination*, p. 42, Macmillan, London; (ii) DIXON, p. 119.

18. (i) Hime, T. W. (1896) *British Medical Journal*, Jenner Centenary Number, 23 May, 1279–89; (ii) Baxby, *Smallpox Vaccine*, pp. 150–64.

19. Laidler, P. W. and Gelfand, M. (1971) *South Africa: Its Medical History 1652–1898*, p. 273, C. Struik, Cape Town.

20. Ceely, R., quoted in Ballard, E. (1868), *Vaccination: a prize essay for the Ladies' Sanitary Association*, p. 256, Longmans Green, London.

21. See SIMON, pp. xiv–xvi.

22. Herrlich, A., Mayr, A., Mahnel, H. and Munz, E. (1963) *Archiv für die gesamte Virusforschung*, 12, 579–99.

23. Miller, G. (1983) *Letters of Edward Jenner*, p.10, Johns Hopkins University Press, Baltimore, MD.

24. Bazin, H. (2000) *The Eradication of Smallpox*, p. 112, Academic Press, London.

25. Dudgeon, J. A. (1963) *British Medical Journal*, 25 May, pp. 1367–72.

26. *British Medical Journal*, 3 May 1902, p. 1102.

27. Bowers, J. Z. (1981) *Bulletin of the History of Medicine*, 55, 32.

28. Copeman, S. M. (1898) *The Lancet*, 21 May, p. 1337.
29. Copeman, S. M. (i) (1891) *Transactions of the International Congress of Hygiene*, vol. ii, p. 325; (ii) (1899) *Vaccination: its natural history and pathology*, pp. 156–61, Macmillan, London.
30. Anderson, E. G. (1899) *Edinburgh Review*, CLXXXIX, 335–68.
31. Hime, T. W. (1896) *British Medical Journal*, Jenner Centenary Number, 23 May, p. 1288.
32. McVail, J. C. (1919) *Half a Century of Smallpox & Vaccination*, Livingstone, Edinburgh.
33. Blaxall, F. R. (1921) *Proceedings of the Royal Society of Medicine*, 27 May, pp. 11–12.
34. Gins, H. A. (1924) *Klinische Wochenschrift*, 3, 634.
35. Arita, I. (1988) in FENNER et al., p. 283.
36. Camus, L. (1909) *Comptes rendus des séances de la Société de Biologie*, 67, 626–9.
37. Shackell, L. F. (1909) *American Journal of Physiology*, 24, 325–40.
38. Arita, I. (1988) in FENNER et al., p. 286.
39. Collier, L. (1957) Tenth World Health Assembly, 7–24 May, *Official Records of the World Health Organisation*, p. 538, quoted in DIXON, pp. 450–51. See also personal communication (1980) quoted in FENNER et al., p. 288.
40. Fenner, F. (1994) in *Control of dual-threat agents: The Vaccines for Peace Programme* (ed. E. Geissler and J. P. Woodall), p. 190, Oxford University Press.
41. ALIBEK, pp. 111–12.

Chapter 13

1. In 1897 Friedrich Löffler and Paul Frosch showed convincingly that the agent of foot-and-mouth disease could pass through fine filters and yet retain the capacity to multiply in infected animals, but their suggested explanation at that time was simply that the agent was a much smaller bacterium than any then known. They did, though, succeed in making an effective vaccine by heating the filtered fluid. See (i) Waterson, A. P. and Wilkinson, L. (1978) *An Introduction to the History of Virology*, pp. 8–13, 30–32, Cambridge University Press; (ii) Rott, R. and Siddell, S. (1998) *Journal of Virology*, 79, pp. 2871–4.
2. See (i) Waterson, and Wilkinson, *History of Virology*, Chapter 3 (ii) Beijerinck, M. W. (1898) *Verhandelingen der Koninklyke Academie van Wettenschappen te Amsterdam*, 65(2), 3–21.
3. Fenner, F. (1988) in FENNER et al., Chapter 2.
4. Buist, J. B. (1887) *Vaccinia and Variola: a study of their life history*, Churchill, London.
5. Smith, G. L., Symons, J. A. and Alcami, A. (1998) *Seminars in Virology*, 8, 409–18.
6. Kerr, J. F. R., Wyllie, A. H. and Currie A. R. (1972) *British Journal of Cancer*, 26, 239–57.
7. Jezek, Z., Szczeniowski, M., Paluku, K. M. and Mutombo. M. (1987) *Journal of Infectious Diseases*, 156, 293–8.
8. Fine, P. E. M., Jezek, Z., Grab, B. and Dixon, H. (1988) *International Journal of Epidemiology*, 17, 643–50.

9. McIntosh, S. K. and McIntosh, R. J. (1981) *American Scientist*, 69, 602–13.
10. (i) Shchelkunov, S. N. et 13 al. (2001) *FEBS letters*, 509, 66–70; (ii) Smith, G. L. and McFadden, G. (2002) *Nature Reviews, Immunology*, 521–7.
11. Letter from Jenner to the Duke of Clarence, 1802, quoted by BARON, vol. 1, p. 522.
12. Gubser, C. and Smith, G. L. (2002) *Journal of General Virology*, 83, 855–72. See also Alfonso, C. L. et 8 al. (2002) *Virology*, 295, 1–9.
13. Baxby, D., Ramyar, H., Hessami, M. and Ghaboosi, B. (1975) *Infection and Immunity*, 11, 617–21. The variola virus used in these experiments was a strain from Tanganyika with properties intermediate between those of variola major and variola minor – see Bedson, H. S., Dumbell, K. R. and Thomas, W. R. G. (1963) *The Lancet*, ii, 1085–8.
14. Downie, A. W. (1939) *Journal of Pathology and Bacteriology*, 48, 361–79. See also *British Journal of Experimental Pathology*, 20, 158–76.
15. Gubser and Smith, *Journal of General Virology*, 83, 855–72, Fig. 3. For earlier results see (i) FENNER et al., pp. 90–94; (ii) Mackett, M. and Archard, L. C. (1979) *Journal of General Virology*, 45, 683–701.
16. Baxby, D. (1981) *Jenner's Smallpox Vaccine*, Chapter 13, Heinemann, London.
17. Mayr, A., Hochstein-Mintzel, V. and Stickl, H. (1975) *Infection*, 3, 6–16, quoted in Smith, G. L., Symons, J. A. and Alcami, A. (1998) *Seminars in Virology*, 8, 409–18.
18. BARON, vol. 2, p. 15.
19. Herrlich, A., Mayr, A., Mahnel, H. and Munz, E. (1963) *Archiv für die gesamte Virusforschung* 12, 579–99.
20. Fenner, F. (1988) in FENNER et al., pp. 90–94.
21. (i) Dudgeon, J. A. (1963) *British Medical Journal*, 25 May, 1367–72; (ii) Baxby, *Smallpox Vaccine*, p.182.
22. Arita, I. (1988) in FENNER et al., p. 582.
23. Ibid.
24. Talk by Inger Damon at a conference 'Twenty-five years on: smallpox revisited' organised by Retroscreen Virology and the Institute of Cell and Molecular Sciences, St Bartholomew's Hospital and the Royal London Hospital, 3 October 2003.

Chapter 14

1. Report submitted by the government of the USSR to the 11th World Health Assembly, 1958, *Official Records of the WHO*, No. 87, Annex 19, pp. 508–12.
2. Henderson, D. A. (1988) in FENNER et al., p. 175.
3. McVail, J. C. (1919) *Half a Century of Small-Pox and Vaccination*, p. 8, Livingstone, Edinburgh.
4. Fenner, F. (1988) in FENNER et al., p. 321.
5. Ibid., p. 324.
6. DIXON, pp. 209–211. There had also been a limited outbreak of *variola minor* in 1901.
7. Fenner, F. (1988) in FENNER et al., p. 324.
8. Howard-Jones, N. (1975) *The scientific background of the International Sanitary Conferences*, 1851–1938, WHO History of International Public Health, No. 1, Geneva.

9. *National Geographic Magazine*, May 2003.
10. Henderson, in FENNER et al., pp. 374–82 and 388–9.
11. Quoted in Henderson, D. A. (1976) *Scientific American*, 235, October, p. 29.
12. Report submitted by the government of the USSR to the 11th World Health Assembly, 1958, *Official Records of the WHO*, No. 87, Annex 19, pp. 508–512.
13. ALIBEK, p. 111–12.
14. Henderson, in FENNER et al., p. 370.
15. Fenner, F. (1988) in FENNER et al., pp. 339–43.
16. Jiang Yutu (1984, 1987) Personal communications quoted in FENNER, pp. 1259–60.
17. Henderson in FENNER et al., Table 9.5, p. 395.
18. Later to become the Center for Disease Control, then the Centers for Disease Control and now the Centers for Disease Control and Prevention.

Chapter 15

1. A full account of the intensified eradication programme is given in Chapters 10–22 of FENNER et al.; a briefer account is in the WHO FINAL REPORT. Accounts of the programme in some individual regions are given in: (i) Basu, R. N., Jezek, Z. and Ward, N. A. (1979) *The eradication of smallpox from India*, WHO History of International Public Health, No. 2, Geneva; (ii) BRILLIANT; (iii) Joarder, A. K., Tarrantola, D. and Tulloch, J. (1980) *The eradication of smallpox from Bangladesh*, WHO Regional Publications: South-East Asia Series, No. 8, New Delhi; (iv) Jezek, Z., Al Aghbari, M., Hatfield, R. and Deria, A. (1981) *Smallpox eradication in Somalia* (ed. J. Tulloch), WHO Regional Office for the Eastern Mediterranean and Somali Democratic Republic, Ministry of Health, Alexandria; (v) Tekeste, Y. et al., (1984) *Smallpox eradication in Ethiopia* (ed. J. Wickett and G. Meiklejohn), WHO Regional Office for Africa, Brazzaville.
2. Henderson, D. A. (1988) in FENNER et al., p. 418.
3. Ibid., Chapter 10.
4. Henderson, D. A. (1976) *Scientific American*, 235, October, pp. 25–33.
5. (i) Foege, W. H., Millar, J. D. and Lane, J. M. (1971) *American Journal of Epidemiology*, 94, 311–15; (ii) Henderson, in FENNER et al., pp. 876–9; (iii) Foege, W. H., Millar, J. D. and Henderson, D. A. (1975) *Bulletin of the WHO*, 52, 209–22.
6. (i) Mack, T. M., Thomas, D. B., Ali, A. and Kahn, M. M. (1972) *American Journal of Epidemiology*, 95, 157–68; (ii) Mack, T. M., Thomas, D. B. and Khan, M. M. (1972) *American Journal of Epidemiology* 95, 169–77.
7. (i) Henderson, in FENNER et al., p. 883; (ii) TUCKER, p. 79.
8. In *northern* Nigeria containment was attempted only in areas in which the mass-vaccination programme had been completed – see Henderson, in FENNER et al., pp. 908–9.
9. (i) WHO FINAL REPORT; (ii) Henderson, in FENNER et al., Chapter 12.
10. Henderson, in FENNER et al., Chapter 13.
11. WHO FINAL REPORT, p. 39.
12. Ibid., pp. 39–42.

13. Ibid., p. 39.
14. Henderson, in FENNER et al., p. 659.
15. Ibid., Chapter 14.
16. Arita, A. (1988) in FENNER et al., pp. 1087–90.
17. TUCKER, p. 84.
18. Henderson, in FENNER et al., Chapter 14, pp. 687–90.
19. Ibid., Chapter 15.
20. BRILLIANT, p. 4.
21. Ibid.
22. (i) Gelfand, H. M. (1966) *American Journal of Public Health*, 56, 1634–51; (ii) Henderson, in FENNER et al., pp. 720–22.
23. TUCKER, p. 96.
24. Shurkin, J. N. (1979, reprinted 2000) *The Invisible Fire*, p. 335, Authors Guild, USA.
25. Henderson, in FENNER et al., pp. 734–5.
26. Ibid., p. 740.
27. Ibid., p. 798.
28. Ibid., p. 744.
29. Ibid., p. 762.
30. For a detailed account of the role of surveillance-containment in the intensified campaign in India (1973–5) see BRILLIANT, pp. 41–68.
31. Ibid., p. 40.
32. Brilliant, L. B. (1978) 'Death for a Killer Disease', *Quest*, May/June.
33. Basu, R. N., Jezek, Z. and Ward, N. A. (1979) *The eradication of smallpox from India*, pp. 79–80, WHO History of International Public Health, No. 2, Geneva.
34. Joarder, A. K., Tarantola, D. and Tulloch, J. (1980) *The eradication of smallpox from Bangladesh*, WHO Regional Publications: South-East Asia Series, No. 8, New Delhi.
35. Henderson, in FENNER et al., p. 820.
36. Shurkin, *Invisible Fire*, p. 354.
37. These figures are based on later surveys for facial pockmarks, as it is reckoned that only about 12 per cent of cases were reported in 1972, and only about 40 per cent in 1973. Hughes, K., Foster, S. O., Tarantola, D., Mehta, H., Tulloch, J. L. and Joarder, A. K. (1980) *International Journal of Epidemiology*, 9, 335–40.
38. Henderson, in FENNER et al., Table 16.16.
39. Joarder, Tarantola, Tulloch, *Eradication of smallpox*.
40. Pankhurst, R. (1965) *Medical History*, 9, 343–55.
41. Henderson, in FENNER et al., Chapter 21.
42. Account by a volunteer, Vincent Radke, quoted in FENNER et al., p. 1012.
43. Tilahun, G., Mondaw, K., Kraushaar, M. D. and Holmberg, M. S. (1972) WHO/SE/72.48, quoted in FENNER et al., p. 1017.
44. Henderson, in FENNER et al., Chapter 22.
45. Ibid., p. 1045.

46. Foster, S. O., El Sid, A. G. H. and Deria, A. (1978) *The Lancet*, 14 October, p. 832.
47. (i) Arita, I. (1988) in FENNER et al., pp. 1097–9; (ii) Shooter, R. A. (Chairman) (1980) *Report of the investigation into the cause of the 1978 Birmingham smallpox occurrence*, HMSO, London.

Chapter 16

1. Churchill, W. S. (1925) quoted in his 1948 book, *The Gathering Storm*, p. 34, Cassell, London.
2. ALIBEK, p. 33.
3. Harris, R. and Paxman, J. (1982) *A Higher Form of Killing*, p. 153, Chatto & Windus, London.
4. ALIBEK, p. 37.
5. Ibid., pp. 30–31.
6. TUCKER, p. 139.
7. Harris, S. (1999) in *Biological and Toxin Weapons* (ed. E. Geissler and J. E. van C. Moon), p. 127, Oxford University Press, Oxford.
8. Harris and Paxman, quoting US National Archives, in *Higher Form*, p. 89.
9. Secretary of War Henry Stimson to Roosevelt, 1942, quoted in ALIBEK, p. 232.
10. Harris and Paxman, *Higher Form*, pp. 216–18.
11. ALIBEK, p. 112.
12. ALIBEK, pp. 112 and 166. It is not clear from Alibek's account what fraction of the stored material consisted of the actual virus.
13. Ibid., p. 43.
14. TUCKER, p. 145.
15. Orent, W. (2000) 'After Anthrax', *The American Prospect*, 8 May, pp. 19–20.
16. TUCKER, pp. 148 and 158.
17. (i) *Now!*, 26 October 1979, quoted in Harris and Paxman, *Higher Form*, p. 220; (ii) ALIBEK, pp. 78–9.
18. Rees, M. (2003) *Our Final Century*, p. 48, Heinemann, London.
19. ALIBEK, p. 140.
20. Preston, R. (2002) *The Demon in the Freezer*, p. 105, Headline, London.
21. ALIBEK, p. 262.
22. (i) Ibid., pp. 259–61; (ii) TUCKER, pp. 158–9.
23. TUCKER, p. 159.
24. ALIBEK, p. 272.
25. TUCKER, p. 202.
26. Orent, W. (19 October 2001) *The Smallpox Wars*, The American Prospect Online.
27. Henderson, D. A. (2003) Lecture on 3 April at a Symposium on Bioterrorism at the Royal Society of Medicine, London.
28. *Washington Post*, 5 November 2002.
29. *Independent*, 22 October 2003, p. 9.
30. (i) Smith, G. L. and McFadden, G. (2002) *Nature Reviews: Immunology*, 2, 521–7;

(ii) WHO Executive Board, 111th Session, provisional agenda item 5.3, 23 December 2002.

31. TUCKER, pp. 166–7.

32. *White House Statement on Destruction of Stocks of Smallpox Virus*, quoted in TUCKER, p. 213.

33. Cello, J., Paul, A. V. and Wimmer, E. (2002) *Science*, 297, 1016–18.

34. CIA panel report entitled *The Darker Bioweapons Future*, 3 November 2003.

35. A cytosine nucleoside analogue which, when phosphorylated, acts as a competitive inhibitor of (and an alternative substrate for) the viral DNA polymerase.

36. See Bradbury, J. (2002) *The Lancet*, 359, 1041, 23 March.

37. Rosengard, A. M., Liu, Y., Zhiping, N. and Jimenez, R. (2002) *Proceedings of the National Academy of Sciences USA*, 99, 8808–13.

38. WHO: EB111/5, 23 December 2002, 'Smallpox eradication: destruction of *Variola virus* stocks'.

39. (i) TUCKER, pp. 36 and 49; (ii) Hopkins, J. W. (1989) *The Eradication of Smallpox: organizational learning and innovation in international Health*, p. 76, Westview Press, Boulder, Colo.

40. See (i) Bartlett, J. et 8 al. (2003) *Clinical Infectious Diseases*, 36, 883–902; (ii) CDC, *Morbidity and Mortality Weekly Report*, 21 February 2003, 52(RR-04), 1–28.

41. Bozzette, et 6 al. (2003) *New England Journal of Medicine*, 348, 416–25.

42. Reuters report, 2 December 2002.

43. Bartlett et al., *Infectious Diseases*.

44. CDC, *Morbidity and Mortality Weekly Report*, 4 April 2003, 52(13), 282–4.

45. Article by D. G. McNeil Jr in *New York Times*, 19 June 2003.

46. Report of conference on the smallpox threat, *British Medical Journal*, 25 October 2003, 327, 948.

47. See, for example, (i) Bozzette et al., *New England Journal of Medicine*; (ii) Article by Oliver Wright in *The Times*, 23 October 2003.

48. (i) *New York Times*, 12 June 2003; (ii) CDC, *Morbidity and Mortality Weekly Reports*, 13, 27 June, 4, 11 July 2003

49. Fine, P. E. M., Jezek, Z., Grab, B. and Dixon, H. (1988) *International Journal of Epidemiology*, 17, 643–50.

50. (i) Hochstein-Mintzel, V., Häninchen, T., Huber, H. Chr. and Stickl, H. (1975) *Zentralblatt für Bakteriologie*, [Orig. A] 230, 283–97; (ii) Mayr, A., Stickl, H., Müller, H. K., Danner, K. and Singer, H. (1978) *Zentralblatt für Bakteriologie*, [Orig. B] 167, 375–90.

51. US Department of Health and Human Services, press release, 25 February 2003. For recent results see Earl, A. L. et 17 al (2004) *Nature* 428, 182–5.

52. Preston, *Demon in the Freezer*, pp. 115–16.

53. (i) Smith, G. L., Mackett, M. and Moss, B. (1983) *Nature*, 302, 490–95; (ii) Smith, G. L., Murphy, B. R. and Moss, B. (1983) *Proceedings of the National Academy of Sciences USA*, 80, 7155–9; (iii) Paoletti, E., Lipinskas, B. R., Samsonoff, C. Mercer, S. and Panicali, D. (1984) *Proceedings of the National Academy of Sciences*

USA, 81, 193–7; (iv) Kieny, M. P. et 8 al. (1984) *Nature*, 312, 163–6.

54. Pastoret, P. P., Brochier, B., Aguilar-Setién, A. and Blancou, J. (1997) in *Veterinary Vaccinology* (ed. P. P. Pastoret, J. Blancou, P. Vannier and C. Verschuren) pp. 616–28, Elsevier, Amsterdam.

55. Hanke, T. (2001) *British Medical Bulletin*, 58, 205–18.

56. (i) Malin, A. S. et 7 al. (2000) *Microbes and Infection*, 2, 1677–85; (ii) Smith, G. L. et 5 al. (1984) *Science*, 224, 397–9; (iii) Moorthy, V. S. et 11 al. (2003) *Vaccine*, 21, 1995–2002.

57. See, for example, Carroll, M. W., Overwijk, W. W., Chamberlain, R. S., Rosenberg, S. A., Moss, B. and Restifo, N. P. (1997) *Vaccine*, 15, 387–94.

58. CDC, *Morbidity and Mortality Weekly Report*, 23 May, 2003/52(20); 471–5.

59. McKenna, M. A. J. (2002) *Atlanta Journal-Constitution*, 27 October.

60. BARON, vol. 2, p. 13.

Index